T0232051

Lecture Notes in Computer Science　　10782

Commenced Publication in 1973
Founding and Former Series Editors:
Gerhard Goos, Juris Hartmanis, and Jan van Leeuwen

More information about this series at http://www.springer.com/series/7407

Arnaud Liefooghe · Manuel López-Ibáñez (Eds.)

Evolutionary Computation in Combinatorial Optimization

18th European Conference, EvoCOP 2018
Parma, Italy, April 4–6, 2018
Proceedings

 Springer

Editors
Arnaud Liefooghe ⓘ
University of Lille
Lille
France

Manuel López-Ibáñez ⓘ
University of Manchester
Manchester
UK

ISSN 0302-9743   ISSN 1611-3349 (electronic)
Lecture Notes in Computer Science
ISBN 978-3-319-77448-0   ISBN 978-3-319-77449-7 (eBook)
https://doi.org/10.1007/978-3-319-77449-7

Library of Congress Control Number: 2018934365

LNCS Sublibrary: SL1 – Theoretical Computer Science and General Issues

Printed on acid-free paper

This Springer imprint is published by the registered company Springer International Publishing AG
part of Springer Nature
The registered company address is: Gewerbestrasse 11, 6330 Cham, Switzerland

Preface

Evolutionary computation techniques and metaheuristics have emerged as methods of choice when tackling challenging problems from combinatorial optimization that appear in various industrial, economic, and scientific domains. In such problems, the goal is to identify high-quality solutions in a large space of discrete variables. Owing to their high complexity and large combinatorics, large and difficult combinatorial optimization problems can typically not be solved to optimality in a reasonable amount of time. This is the reason why decision-makers and analysts have to rely on heuristic algorithms. Among them, metaheuristics comprise a whole family of optimization approaches that include evolutionary algorithms and other nature-inspired approaches such as ant colony and particle swarm optimization, as well as advanced local search methods such as simulated annealing and tabu search. These search paradigms rely on stochastic operators, and constitute general-purpose methodologies that guide the design of heuristic algorithms for solving a given optimization problem. At the crossroads of computer science, discrete mathematics, operations research and decision support, computational intelligence, and machine learning, the successful use of evolutionary computation for combinatorial optimization is the main subject of these proceedings.

This volume contains the proceedings of EvoCOP 2018, the 18th European Conference on Evolutionary Computation in Combinatorial Optimization. The conference was held in Parma, Italy, during April 4–6, 2018. The EvoCOP conference series started in 2001, with the first workshop specifically devoted to evolutionary computation in combinatorial optimization. It became an annual conference in 2004. EvoCOP 2018 was organized together with EuroGP (the 21st European Conference on Genetic Programming), EvoMUSART (the 7th International Conference on Computational Intelligence in Music, Sound, Art and Design), and EvoApplications (the 21st European Conference on the Applications of Evolutionary Computation, formerly known as EvoWorkshops), in a joint event collectively known as EvoStar 2018. Previous EvoCOP proceedings were published by Springer in the *Lecture Notes in Computer Science* series (LNCS volumes 2037, 2279, 2611, 3004, 3448, 3906, 4446, 4972, 5482, 6022, 6622, 7245, 7832, 8600, 9026, 9595, and 10197). The table on the next page reports the statistics for each of the previous conference.

This year, 12 out of 37 papers were accepted after a rigorous double-blind process, resulting in a 32% acceptance rate. We would like to acknowledge the quality and timeliness of our Program Committee members' work. Decisions considered both the reviewers' report and the evaluation of the program chairs. The 12 accepted papers cover a wide spectrum of topics, ranging from the foundations of evolutionary computation algorithms and other search heuristics, to their accurate design and application to both single- and multi-objective combinatorial optimization problems. Fundamental and methodological aspects deal with runtime analysis, the structural properties of fitness landscapes, the study of metaheuristics core components, the clever design

EvoCOP	LNCS vol.	Submitted	Accepted	Acceptance (%)
2018	10782	37	12	32.4
2017	10197	39	16	41.0
2016	9595	44	17	38.6
2015	9026	46	19	41.3
2014	8600	42	20	47.6
2013	7832	50	23	46.0
2012	7245	48	22	45.8
2011	6622	42	22	52.4
2010	6022	69	24	34.8
2009	5482	53	21	39.6
2008	4972	69	24	34.8
2007	4446	81	21	25.9
2006	3906	77	24	31.2
2005	3448	66	24	36.4
2004	3004	86	23	26.7
2003	2611	39	19	48.7
2002	2279	32	18	56.3
2001	2037	31	23	74.2

of their search principles, and their careful selection and configuration by means of automatic algorithm configuration and hyper-heuristics. Applications cover conventional academic domains such as NK landscapes, binary quadratic programming, traveling salesman, vehicle routing, or scheduling problems, and also include real-world domains in clustering, commercial districting, and winner determination. It is our hope that the wide range of topics covered in this volume of EvoCOP proceedings reflects the current state of research in the fields of evolutionary computation and combinatorial optimization.

We would like to express our appreciation to the various persons and institutions involved in making EvoCOP 2018 a successful event. Firstly, we thank the local organization team, led by Stefano Cagnoni and Monica Mordonini, from the University of Parma, Italy. We extend our gratitude to Marc Schoenauer from Inria Saclay, France, for his continued assistance in providing the MyReview conference management system, and to Pablo García-Sánchez from the University of Cádiz, Spain, for the EvoStar publicity and website. Thanks are also due to our EvoStar coordinators Anna I Esparcia-Alcázar, from Universitat Politècnica de València, Spain, and Jennifer Willies, as well as to the SPECIES (Society for the Promotion of Evolutionary Computation in Europe and its Surroundings) executive board, including Marc Schoenauer (President), Anna I Esparcia-Alcázar (Secretary and Vice-President), and Wolfgang Banzhaf (Treasurer). We finally wish to thank our prominent keynote speakers, Una-May O'Reilly (MIT Computer Science and Artificial Intelligence Laboratory, USA) and Penousal Machado (Computational Design and Visualization Lab at the University of Coimbra, Portugal).

Special thanks also to Christian Blum, Francisco Chicano, Carlos Cotta, Peter Cowling, Jens Gottlieb, Jin-Kao Hao, Bin Hu, Jano van Hemert, Peter Merz, Martin Middendorf, Gabriela Ochoa, and Günther R. Raidl for their hard work and dedication during past editions of EvoCOP, making this one of the reference international events in evolutionary computation and metaheuristics.

April 2018   Arnaud Liefooghe
                                                    Manuel López-Ibáñez

Organization

EvoCOP 2018 was organized jointly with EuroGP 2018, EvoMUSART 2018, and EvoApplications 2018.

Organizing Committee

Program Chairs

Arnaud Liefooghe	University of Lille, France
Manuel López-Ibáñez	University of Manchester, UK

Local Organization

Stefano Cagnoni	University of Parma, Italy
Monica Mordonini	University of Parma, Italy

Publicity Chair

Pablo García-Sánchez	University of Cádiz, Spain

EvoCOP Steering Committee

Christian Blum	Artificial Intelligence Research Institute (IIIA-CSIC), Bellaterra, Spain
Francisco Chicano	University of Málaga, Spain
Carlos Cotta	University of Málaga, Spain
Peter Cowling	University of York, UK
Jens Gottlieb	SAP AG, Germany
Jin-Kao Hao	University of Angers, France
Bin Hu	AIT Austrian Institute of Technology, Austria
Jano van Hemert	Optos, UK
Peter Merz	Hannover University of Applied Sciences and Arts, Germany
Martin Middendorf	University of Leipzig, Germany
Gabriela Ochoa	University of Stirling, UK
Günther Raidl	Vienna University of Technology, Austria

Society for the Promotion of Evolutionary Computation in Europe and its Surroundings (SPECIES)

Marc Schoenauer	President
Anna I Esparcia-Alcázar	Secretary and Vice-President
Wolfgang Banzhaf	Treasurer
Jennifer Willies	EvoStar coordinator with Anna I Esparcia-Alcázar

Program Committee

Adnan Acan	Eastern Mediterranean University, Turkey
Hernán Aguirre	Shinshu University, Japan
Enrique Alba	University of Málaga, Spain
Richard Allmendinger	University of Manchester, UK
Thomas Bartz-Beielstein	Cologne University of Applied Sciences, Germany
Benjamin Biesinger	Austrian Institute of Technology, Austria
Christian Blum	Artificial Intelligence Research Institute (IIIA-CSIC), Bellaterra, Spain
Sandy Brownlee	University of Stirling, UK
Pedro Castillo	University of Granada, Spain
Francisco Chicano	University of Málaga, Spain
Carlos Coello Coello	CINVESTAV-IPN, Mexico
Bilel Derbel	University of Lille, France
Luca DiGaspero	University of Udine, Italy
Karl Doerner	Johannes Kepler University Linz, Austria
Benjamin Doerr	LIX-Ecole Polytechnique, France
Carola Doerr	Université Pierre et Marie Curie, France
Paola Festa	Universitá di Napoli Federico II, Italy
Bernd Freisleben	University of Marburg, Germany
Carlos Garcia-Martinez	University of Córdoba, Spain
Adrien Goeffon	University of Angers, France
Jens Gottlieb	SAP, Germany
Walter Gutjahr	University of Vienna, Austria
Said Hanafi	University of Valenciennes, France
Jin-Kao Hao	University of Angers, France
Emma Hart	Edinburgh Napier University, UK
Geir Hasle	SINTEF Applied Mathematics, Norway
Bin Hu	Austrian Institute of Technology, Austria
Andrzej Jaszkiewicz	Poznan University of Technology, Poland
Istvan Juhos	University of Szeged, Hungary
Graham Kendall	University of Nottingham, UK
Ahmed Kheiri	Lancaster University, UK
Mario Koeppen	Kyushu Institute of Technology, Japan
Timo Koetzing	Hasso Plattner Institute, Germany
Frederic Lardeux	University of Angers, France

Rhyd Lewis   Cardiff University, UK
Jose Antonio Lozano   University of the Basque Country, Spain
Gabriel Luque   University of Málaga, Spain
David Meignan   University of Osnabruck, Germany
Juan Julian Merelo   University of Granada, Spain
Krzysztof Michalak   University of Economics, Wroclaw, Poland
Martin Middendorf   University of Leipzig, Germany
Christine L. Mumford   Cardiff University, UK
Nysret Musliu   Vienna University of Technology, Austria
Gabriela Ochoa   University of Stirling, UK
Beatrice   Brock University, Canada
  Ombuki-Berman
Luis Paquete   University of Coimbra, Portugal
Mario Pavone   University of Catania, Italy
Paola Pellegrini   French Institute of Science and Technology for Transport,
                            France
Francisco J. Pereira   Universidade de Coimbra, Portugal
Jakob Puchinger   SystemX-Centrale Supélec, France
Günther Raidl   Vienna University of Technology, Austria
Maria Cristina Riff   Universidad Técnica Federico Santa María, Chile
Eduardo   CINVESTAV - Tamaulipas, Mexico
  Rodriguez-Tello
Andrea Roli   Università di Bologna, Italy
Peter Ross   Edinburgh Napier University, UK
Frederic Saubion   University of Angers, France
Patrick Siarry   University of Paris 12, France
Kevin Sim   Edinburgh Napier University, UK
Jim Smith   University of the West of England, UK
Giovanni Squillero   Politecnico di Torino, Italy
Thomas Stuetzle   Université Libre de Bruxelles, Belgium
El-ghazali Talbi   University of Lille, France
Sara Tari   University of Angers, France
Renato Tinós   University of Sao Paulo, Brasil
Nadarajen Veerapen   University of Stirling, UK
Sébastien Verel   Université du Littoral Cote d'Opale, France
Bing Xue   Victoria University of Wellington, New Zealand
Takeshi Yamada   NTT Communication Science Laboratories, Japan

Contents

Better Runtime Guarantees
via Stochastic Domination

Benjamin Doerr$^{(\boxtimes)}$

Ecole Polytechnique, Palaiseau, France

Abstract. Apart from few exceptions, the mathematical runtime analysis of evolutionary algorithms is mostly concerned with expected runtimes. In this work, we argue that stochastic domination is a notion that should be used more frequently in this area. Stochastic domination allows to formulate much more informative performance guarantees than the expectation alone, it allows to decouple the algorithm analysis into the true algorithmic part of detecting a domination statement and probability theoretic part of deriving the desired probabilistic guarantees from this statement, and it allows simpler and more natural proofs.

As particular results, we prove a fitness level theorem which shows that the runtime is dominated by a sum of independent geometric random variables, we prove tail bounds for several classic problems, and we give a short and natural proof for Witt's result that the runtime of any (μ, p) mutation-based algorithm on any function with unique optimum is subdominated by the runtime of a variant of the $(1 + 1)$ EA on the ONEMAX function.

1 Introduction

The analysis of evolutionary algorithms via mathematical means is an established part of evolutionary computation research. The subarea of *runtime analysis* aims at giving proven performance guarantees on the time an evolutionary algorithm takes to find optimal or near-optimal solutions. Traditionally, this area produces estimates for the expected runtime, which are occasionally augmented by tail bounds. A justification for this was that for already very simply evolutionary algorithms and optimization problems, the stochastic processes arising from running this algorithm on this problem are so complicated that any more detailed analysis is infeasible. See the analysis how the $(1+1)$ evolutionary algorithm optimizes linear functions $x \mapsto a_1 x_1 + \cdots + a_n x_n$ in [1] for an example for this complexity.

In this work, we shall argue that the restriction to expectations is only partially justified and propose stochastic domination as an alternative. It is clear that the precise distribution of the runtime of an evolutionary algorithms in all cases apart from very trivial ones is out of reach. Finding the precise distribution is maybe not even an interesting target because most likely already the result will be too complicated to be useful. What would be very useful is

© Springer International Publishing AG, part of Springer Nature 2018
A. Liefooghe and M. López-Ibáñez (Eds.): EvoCOP 2018, LNCS 10782, pp. 1–17, 2018.
https://doi.org/10.1007/978-3-319-77449-7_1

a possibly not absolutely tight upper bound-type statement that concerns the whole distribution of the runtime.

One way to formalize such statement is via the notion of *stochastic domination*. We say that a real-valued random variable Y stochastically dominates another one X if and only if for each $\lambda \in \mathbb{R}$, we have $\Pr[X \le \lambda] \ge \Pr[Y \le \lambda]$. If X and Y describe the runtimes of two algorithms A and B, then this domination statement is a very strong way of saying that algorithm A is at least as fast as B. See Sect. 2 for a more detailed discussion of the implication of such a domination statement.

In this work, we shall give three main arguments for a more frequent use of domination arguments in runtime analysis.

Stochastic domination is often easy to show. Surprisingly, despite being a much stronger type of assertion, stochastic domination statements are often easy to obtain. The reason is that many of the classic proofs implicitly contain all necessary information, they only fail to formulate the result as a domination statement. As an example, we prove a natural domination version of the classic fitness level method. In analogy to the classic result, which translates pessimistic improvement probabilities $p_1, \ldots, p_{m-1}$ into an expected runtime estimate $E[T] \le \sum_{i=1}^{m-1} \frac{1}{p_i}$, we show that under the same assumptions the runtime T is dominated by the sum of independent geometric random variables with success probabilities $p_1, \ldots, p_{m-1}$. This statement implies the classic result, but also implies tail bound for the runtime via Chernoff bounds for geometric random variables. We note that, while our extension is very natural, the proof is not totally obvious, which might explain why previous works were restricted to the expectation version.

Stochastic domination allows to separate the core algorithm analysis and the probability theoretic derivation of probabilistic runtime statements. The reason why stochastic domination statements are often easy to obtain is that they are close to the actions of the algorithms. When we are waiting for the algorithm to succeed in performing a particular action, then it is a geometric distribution that describes this waiting time. To obtain such a statement, we need to understand how the algorithm reaches a particular goal. This requires some elementary probability theory and discrete mathematics, but usually no greater expertise in probability theory. We give several examples, mostly using the fitness level method, but also for the single-source shortest paths problem, where the fitness level method is not suitable to give the best known results.

Once we have a domination statement formulated, e.g., that the runtime is dominated by a sum of independent geometric distributions, then deeper probability theoretic arguments like Chernoff-type tail bounds come into play. This part of the analysis is independent of the algorithm and only relies on the domination statement. Exemplarily, we derive tail bounds for the runtime of the $(\mu + 1)$ EA on ONEMAX, the $(1 + 1)$ EA for sorting, and the multi-criteria $(1 + 1)$ EA for the single-source shortest path problem.

That these two stages of the analysis are of a different nature is also visible in the history of runtime analysis. As discussed in the previous paragraph, the classic fitness level method essentially contains all ingredients to formulate a runtime domination statement. However, the mathematical tools to analyze sums of independent geometric random variables were developed much later (and this development is still ongoing).

This historical note also shows that from the viewpoint of research organization, it would have been profitable if previous works would have been formulated in terms of domination statements. This might have spurred a faster development of suitable Chernoff bounds and, in any case, it would have made it easier to apply the recently found Chernoff bounds (see Theorems 3 to 5) to these algorithmic problems.

Stochastic domination often leads to more natural and shorter proofs. To demonstrate this in an application substantially different from the previous ones, we regard the classic lower-bound result which, in simple words, states that ONE-MAX is the easiest fitness function for many mutation-based algorithms. This statement, of course, should be formulated via stochastic domination (and this has indeed been done in previous work). However, as we shall argue in Sect. 5, we also have the statement that when comparing the optimization of ONE-MAX and some other function with unique optimum, then the distance of the current-best solution from the optimum for the general function dominates this distance for the ONEMAX function. This natural statement immediately implies the domination relation between the runtimes. We make this precise for the current-strongest ONEMAX-is-easiest result [2]. This will shorten the previous, two-and-a-half pages long complicated proof to an intuitive proof of less than a page.

Related work. The use of stochastic domination is not totally new in the theory of evolutionary computation, however, the results so far appear rather sporadic than systematic. In [3], the precise analysis of the expected runtime of the $(1 + 1)$ EA on the LEADINGONES function [4] was transformed into a domination statement, which was then used to derive a tail bound, which again was necessary to obtain results in the fixed budget perspective [5].

Before that, in [6], different evolutionary algorithms were compared by arguing that at all times the fitness in one algorithm dominates the fitness in the other algorithm. This was done, however, without explicitly appealing to the notion of stochastic domination. Consequently, the tools presented in this work were not available. Still, this might be the first work in the theory of evolutionary computation which uses stochastic domination to phrase and prove results.

What comes closest to this work is the paper [7], which also tries to establish runtime analysis beyond expectations. The notion proposed in [7], called *probable computational time* $L(\delta)$, is the smallest time T such that the algorithm under investigation within the first T fitness evaluations finds an optimal solution with probability at least $1 - \delta$. In some sense, this notion can be seen as an inverse of the classic tail bound language, which makes statements of the type that the probability that the algorithm needs more than T iterations, is at most some function in T, which is often has an exponential tail. To prove results on the probable computational time, [7] implicitly use (but do not prove) the result that the fitness level method gives a stochastic domination by a sum of independent geometric random variables (which we prove in Sect. 3). When using this result to obtain bounds on the probable computational time, [7] suffer from the fact that at that time, good tail bounds for sums of geometric random variables with different success probabilities where not yet available (these appeared only in [8, 9]). For this reason, they had to prove an own tail bound, which unfortunately gives a non-trivial tail probability only for values above twice the expectation. With this restriction, [7] prove tail bounds for the runtimes of the algorithms RLS, $(1+1)$ EA, $(\mu+1)$ EA, MMAS* and binary PSO on the optimization problems ONEMAX and LEADINGONES.

2   Stochastic Domination

In this section, recall the definition of *stochastic domination* and collect a few known properties of this notion. For an extensive treatment of various forms of stochastic orders, we refer to [10].

Definition 1 (Stochastic domination). *Let X and Y be random variables not necessarily defined on the same probability space. We say that Y stochastically dominates X, written as $X \preceq Y$, if for all $\lambda \in \mathbb{R}$ we have $\Pr[X \leq \lambda] \geq \Pr[Y \leq \lambda]$.*

If Y dominates X, then the cumulative distribution function of Y is pointwise not larger than the one of X. The definition of domination is equivalent to "$\forall \lambda \in \mathbb{R} : \Pr[X \geq \lambda] \leq \Pr[Y \geq \lambda]$", which shows more clearly why we feel that Y is at least as large as X.

Concerning nomenclature, we remark that some research communities in addition require that the inequality is strict for at least one value of λ. Hence, intuitively speaking, Y is strictly larger than X. From the mathematical perspective, this appears not very practical, and from the computer science perspective it is not clear what can be gained from this alternative definition. Consequently, our definition above is more common in computer science (though, e.g., [7] also use the alternative definition).

One advantage of comparing two distributions via the notion of domination is that this makes a statement over the whole domain of the distributions, including "rare events" on the tails. If the runtime T_A of some algorithm A is dominated by the runtime T_B of algorithm B, then not only A is better than B in average, but also exceptionally large runtimes occur less frequent when using A.

A second advantage is that domination is invariant under monotonic rescaling. Imagine that running an algorithm for time t incurs some cost $c(t)$. We may clearly assume that c is monotonically increasing, that is, that $t_1 \leq t_2$ implies $c(t_1) \leq c(t_2)$. Then $T_A \preceq T_B$ implies $c(T_A) \preceq c(T_B)$. Hence changing the cost measure does not change our feeling that algorithm A is better than B. Note that this is different for expectations. We may well have $E[T_A] < E[T_B]$, but $E[c(T_A)] > E[c(T_B)]$. We collect a few useful properties of stochastic domination.

Lemma 1. *If $X \preceq Y$, then $E[X] \leq E[Y]$.*

Lemma 2. *The following two conditions are equivalent.*

(i) $X \preceq Y$.
(ii) For all monotonically increasing functions $f : \mathbb{R} \to \mathbb{R}$, we have

$$E[f(X)] \leq E[f(Y)].$$

The following non-trivial lemma will be needed in our proof of the extended fitness level theorem. It was proven in slightly different forms in [11] and [12] with the additional assumption that for all $i \in [1..n]$, the variable X_i^* is independent of $X_1, \ldots, X_{i-1}$, but a short reflection of the proof reveals that this can be assumed with out loss of generality, so it is not required in the statement of the result. Also, as a reviewer correctly remarked, the restriction to integral random variables can easily be relaxed to arbitrary discrete random variables.

Lemma 3. *Let $X_1, \ldots, X_n$ be arbitrary integral random variables. Let $X_1^*, \ldots, X_n^*$ be random variables that are mutually independent. Assume that for all $i \in [1..n]$ and all $x_1, \ldots, x_{i-1} \in \mathbb{Z}$ with $\Pr[X_1 = x_1, \ldots, X_{i-1} = x_{i-1}] > 0$, we have*

$$\Pr[X_i \geq k \mid X_1 = x_1, \ldots, X_{i-1} = x_{i-1}] \leq \Pr[X_i^* \geq k]$$

for all $k \in \mathbb{Z}$, that is, X_i^* dominates $(X_i \mid X_1 = x_1, \ldots, X_{i-1} = x_{i-1})$. Then for all $k \in \mathbb{Z}$, we have

$$\Pr\left[\sum_{i=1}^n X_i \geq k\right] \leq \Pr\left[\sum_{i=1}^n X_i^* \geq k\right],$$

that is, $\sum_{i=1}^n X_i^*$ dominates $\sum_{i=1}^n X_i$.

Stochastic domination is tightly connected to *coupling*. Coupling is an analysis technique that consists of defining two unrelated random variables over the same probability space to ease comparing them. Let X and Y be two random variables not necessarily defined over the same probability space. We say that $(\tilde{X}, \tilde{Y})$ is a *coupling* of (X, Y) if $\tilde{X}$ and $\tilde{Y}$ are defined over a common probability space and if X and X' as well as Y and Y' are identically distributed. This definition itself is very weak. (X, Y) have many couplings and most of them are not interesting. So the art of coupling as a proof and analysis technique is to find a coupling of (X, Y) that allows to derive some useful information. This is often problem-specific, however, also the following general result in known. It in particular allows to couple dominating random variables. We shall use it for this purpose in Sect. 5.

Theorem 1. *Let X and Y be random variables. Then the following two statements are equivalent.*

(i) $X \preceq Y$.
(ii) There is a coupling $(\tilde{X}, \tilde{Y})$ of (X, Y) such that $\tilde{X} \leq \tilde{Y}$.

3   Domination-Based Fitness Level Method

In this section, we prove a fitness level theorem that gives a domination statement and we apply it to two classic problems. The fitness level method, invented by Wegener [13], is one of the most successful early analysis methods in the theory of evolutionary computation. If builds on the idea of partitioning the search space into levels A_i, $i = 1, \ldots, m$, which contain search points of strictly increasing fitness. We then try to show a lower bound p_i for the probability that, given that the current-best search point is in A_i, we generate in one iteration a search point in a higher level. From this data, the fitness level theorem gives an estimate of $E[T] \leq \sum_{i=1}^{m-1} \frac{1}{p_i}$ for the time T to find a search point in the highest level (which traditionally is assumed to contain only optimal solutions, though this is not necessary).

We shall now show that under the same assumptions, a much stronger statement is valid, namely that the runtime T is dominated by $\sum_{i=1}^{m-1} \mathrm{Geom}(p_i)$, that is, a sum independent random variables following geometric distributions with success probabilities p_i. This result appears to be very natural and was used without proof in [7], yet its proof requires the non-trivial Lemma 3.

Theorem 2 (Domination version of the fitness level method). *Consider an elitist evolutionary algorithm $\mathcal{A}$ maximizing a function $f : \Omega \to \mathbb{R}$. Let $A_1, \ldots, A_m$ be a partition of Ω such that for all $i, j \in [1..m]$ with $i < j$ and all $x \in A_i$, $y \in A_j$, we have $f(x) < f(y)$. Set $A_{\geq i} := A_i \cup \cdots \cup A_m$. Let $p_1, \ldots, p_{m-1}$ be such that for all $i \in [1..m-1]$ we have that if the best search point in the current parent population is contained in A_i, then independently of the past with probability at least p_i the next parent population contains a search point in $A_{\geq i+1}$.*

Denote by T the (random) number of iterations $\mathcal{A}$ takes to generate a search point in A_m. Then

$$T \preceq \sum_{i=1}^{m-1} \mathrm{Geom}(p_i),$$

where this sum is to be understood as a sum of independent geometric distributions.

Proof. Consider a run of the algorithm $\mathcal{A}$. For all $i \in [1..m]$, let T_i be the first time (iteration) when $\mathcal{A}$ has generated a search point in $A_{\geq i}$. Then $T = T_m = \sum_{i=1}^{m-1}(T_{i+1} - T_i)$. By assumption, $T_{i+1} - T_i$ is dominated by a geometric random variable with parameter p_i regardless what happened before time T_i. Consequently, Lemma 3 gives the claim. □

By Lemma 1, the expected runtime in Theorem 2 satisfies $E[T] \leq \sum_{i=1}^{m-1} \frac{1}{p_i}$, which is the common version of the fitness level theorem [13]. However, by using tail bounds for sums of independent geometric random variables, we also obtain runtime bounds that hold with high probability. This was first proposed in [7], but did not give very convincing results due to the lack of good tail bounds at that time. We briefly present the tail bounds known by now and then give a few examples how to use them together with the new fitness level theorem.

Theorem 3. *Let* $X_1, \ldots, X_n$ *be independent geometric random variables with success probabilities* $p_1, \ldots, p_n$. *Let* $p_{\min} := \min\{p_i \mid i \in [1..n]\}$. *Let* $X := \sum_{i=1}^n X_i$ *and* $\mu = E[X] = \sum_{i=1}^n \frac{1}{p_i}$.

(i) For all $\delta \geq 0$,

$$\Pr[X \geq (1+\delta)\mu] \leq \frac{1}{1+\delta}(1 - p_{\min})^{\mu(\delta - \ln(1+\delta))} \tag{1}$$

$$\leq \exp(-p_{\min}\mu(\delta - \ln(1+\delta))) \tag{2}$$

$$\leq \left(1 + \frac{\delta\mu p_{\min}}{n}\right)^n \exp(-\delta\mu p_{\min}) \tag{3}$$

$$\leq \exp\left(-\frac{(\delta\mu p_{\min})^2}{2n(1 + \frac{\delta\mu p_{\min}}{n})}\right). \tag{4}$$

(ii) For all $0 \leq \delta \leq 1$,

$$\Pr[X \leq (1-\delta)\mu] \leq (1-\delta)^{p_{\min}\mu}\exp(-\delta p_{\min}\mu) \tag{5}$$

$$\leq \exp\left(-\frac{\delta^2\mu p_{\min}}{2 - \frac{4}{3}\delta}\right) \tag{6}$$

$$\leq \exp(-\tfrac{1}{2}\delta^2\mu p_{\min}). \tag{7}$$

Estimates (1) and (2) are from [14], bound (3) is from [15], and (4) follows from the previous by standard estimates. The lower tail bound (5) is from [14]. It implies (6) via standard estimates. Inequality (7) appeared in [15].

It is surprising that none of these useful bounds appeared in a reviewed journal. For the case that all geometric random variables have the same success probability p, the bound

$$\Pr[X \geq (1+\delta)\mu] \leq \exp\left(-\frac{\delta^2}{2}\frac{n-1}{1+\delta}\right) \tag{8}$$

appeared in [16].

The bounds of Theorem 3 allow the geometric random variables to have different success probabilities, however, the tail probability depends only on the smallest of them. This is partially justified by the fact that the corresponding geometric random variable has the largest variance, and thus might be most detrimental to the desired strong concentration. If the success probabilities vary significantly, however, then this approach gives overly pessimistic tail bounds. Witt [8] proves the following result, which can lead to stronger estimates.

Theorem 4. *Let* $X_1, \ldots, X_n$ *be independent geometric random variables with success probabilities* $p_1, \ldots, p_n$. *Let* $X = \sum_{i=1}^n X_i$, $s = \sum_{i=1}^n (\frac{1}{p_i})^2$, *and* $p_{\min} := \min\{p_i \mid i \in [1..n]\}$. *Then for all* $\lambda \geq 0$,

$$\Pr[X \geq E[X] + \lambda] \leq \exp\left(-\frac{1}{4}\min\left\{\frac{\lambda^2}{s}, \lambda p_{\min}\right\}\right), \tag{9}$$

$$\Pr[X \leq E[X] - \lambda] \leq \exp\left(-\frac{\lambda^2}{2s}\right). \tag{10}$$

As we shall see, we often encounter sums of independent geometrically distributed random variables $X_1, \ldots, X_n$ with success probabilities p_i proportional to i. For this case, the following result from [9] gives stronger tail bounds than the previous result. Recall that the harmonic number H_n is defined by $H_n = \sum_{i=1}^n \frac{1}{i}$.

Theorem 5. *Let $X_1, \ldots, X_n$ be independent geometric random variables with success probabilities $p_1, \ldots, p_n$. Assume that there is a number $C \leq 1$ such that $p_i \geq C\frac{i}{n}$ for all $i \in [1..n]$. Let $X = \sum_{i=1}^n X_i$. Then*

$$E[X] \leq \frac{1}{C}nH_n \leq \frac{1}{C}n(1 + \ln n), \tag{11}$$

$$\Pr[X \geq (1 + \delta)\frac{1}{C}n\ln n] \leq n^{-\delta} \text{ for all } \delta \geq 0. \tag{12}$$

Applications of the Fitness Level Theorem

We now present a few examples where our fitness level theorem gives more details about the distribution of the runtime and where this, together with the just presented tail bounds, allows to obtain tail bounds for the runtime.

We note that so far only very few results exist that give detailed information about the runtime distribution. A well-known triviality is that the runtime of the *randomized local search* (RSH) heuristic on the ONEMAX function strongly related to the coupon collector process. In particular, its runtime is dominated by the coupon collecting time, which is $\sum_{i=1}^n \text{Geom}(\frac{i}{n})$. With arguments similar to those in [17], this can be sharpened slightly to the result that the runtime is dominated by $\sum_{i=1}^{\lceil n/2 \rceil} \text{Geom}(\frac{i}{n})$.

Apart from this, we are only aware of the result in [3], which shows that the runtime T of the $(1+1)$ EA on the LEADINGONES function is $\sum_{i=0}^{n-1} X_i A_i$, where the X_i are uniformly distributed on $\{0, 1\}$, $A_i \sim \text{Geom}(\frac{1}{n}(1 - \frac{1}{n})^i)$, and all these random variables are mutually independent (in Lemma 5 of [3], the range of the sum starts at 1, but this appears to be a typo).

ONEMAX. How the $(1+1)$ EA optimizes the ONEMAX test function ONEMAX : $\{0,1\}^n \rightarrow \mathbb{R}; (x_1, \ldots, x_n) \mapsto \sum_{i=1}^n x_i$ is one of the first results in runtime analysis. Using the fitness level method with the levels $A_i := \{x \in \{0,1\}^n \mid \text{ONEMAX}(x) = i\}$, $i = 0, 1, \ldots, n$, one easily obtains that the expectation of the runtime T is at most enH_n. For this, it suffices to estimate $p_i \geq \frac{n-i}{n}(1 - \frac{1}{n})^{n-1} \geq \frac{n-i}{en}$, see [1] for the details. By Theorems 2 and 5, we also obtain

$$T \preceq \sum_{i=1}^n \text{Geom}\left(\frac{i}{en}\right),$$

$$\Pr[T \geq (1 + \delta)en\ln n] \leq n^{-\delta} \text{ for all } \delta \geq 0.$$

The domination result, while not very deep, appears to be new, whereas the tail bound was proven before via multiplicative drift analysis [18].

Things become more interesting (and truly new) when regarding the $(1 + \lambda)$ EA instead of the $(1+1)$ EA. We denote by $d(x) = n - \text{ONEMAX}(x)$ the distance of x to the optimum. Let $t = \lfloor \frac{\ln(\lambda)-1}{2\ln\ln\lambda} \rfloor$. We partition the lowest $L = \lfloor n - \frac{n}{\ln\lambda} \rfloor$ fitness levels (that is, the search points with $\text{ONEMAX}(x) < L$) into $\lceil \frac{L}{t} \rceil$ sets each spanning at most t consecutive fitness levels. In [19], it was shown that the probability to leave such a set in one iteration is at least $p_0 \geq (1 - \frac{1}{e})$. The remaining search points are partitioned into sets of points having equal fitness, that is, $A_i = \{x \mid \text{ONEMAX}(x) = i\}$, $i = L, \ldots, n$. For these levels, the probability to leave a set in one iteration is at least

$$p_i \geq 1 - \left(1 - \left(1 - \frac{1}{n}\right)^{n-1} \frac{n-i}{n}\right)^{\lambda} \geq 1 - \left(1 - \frac{n-i}{en}\right)^{\lambda}.$$

For $i \leq n - \frac{en}{\lambda}$, this is at least $p_i \geq 1 - \frac{1}{e}$ by the well-known estimate $1 + r \leq e^r$ valid for all $r \in \mathbb{R}$. For $i > n - \frac{en}{\lambda}$, we use an inverse version of Bernoulli's inequality to estimate $p_i \geq 1 - (1 - \frac{\lambda}{en}(n-i) + \frac{1}{2}(\frac{\lambda}{en}(n-i))^2) \geq \frac{1}{2}\frac{\lambda}{en}(n-i)$. Using the abbreviation $T_0 := \lceil \frac{L}{t} \rceil + \lceil \frac{n}{\ln\lambda} \rceil - \lceil \frac{en}{\lambda} \rceil + 1$, our fitness level theorem gives the following domination bound for the number T of *iterations* until the optimum is found.

$$T \preceq \sum_{i=1}^{T_0} \text{Geom}(1 - \tfrac{1}{e}) + \sum_{i=1}^{\lceil \frac{en}{\lambda} - 1 \rceil} \text{Geom}(\tfrac{1}{2}\tfrac{\lambda i}{en}).$$

To avoid uninteresting case distinctions, we now concentrate on the more interesting case that $\lambda = \omega(1)$ and $\lambda = n^{O(1)}$. In this case, the above domination statement immediately gives

$$E[T] \leq \frac{e}{e-1} T_0 + 2e \frac{n \ln(\lceil \frac{en}{\lambda} \rceil)}{\lambda} = (1 + o(1)) \left(\frac{2e}{e-1} \frac{n \ln\ln\lambda}{\ln\lambda} + 2e \frac{n \ln(n)}{\lambda}\right),$$

which is the result of [19] with explicit constants. With Theorems 3 and 5 we also obtain the (new) tail bound

$$\Pr[T \geq E[T] + K] \leq ((1 + \tfrac{(e-1)K}{2eT_0}) \exp(-\tfrac{(e-1)K}{2eT_0}))^{\frac{e}{e-1}T_0} + (\tfrac{\lambda}{2en})^{K\lambda/2en \ln(en/\lambda)}.$$

We note that this expression is not very easy to parse. This is caused by its generality, that is, that is covers the two cases that each of the two parts of the expected runtime is the dominant one. When we have more information about the size of λ, both the term simplifies and a simpler estimate could be used in the first place.

Sorting. One of the first combinatorial problems regarded in the theory of evolutionary computation is how a combinatorial $(1 + 1)$ EA sorts an array of length n. One of several setups regarded in [20] is modelling the sorting problem as the minimization of the number of inversions in the array. We assume that the $(1 + 1)$ EA mutates an array by first determining a number k according to a Poisson distribution with parameter $\lambda = 1$ and then performing $k + 1$ random exchanges (to ease the presentation, we do not use the jump operations also employed in [20], but it is easy to see that this does not significantly change things). It is easy to see that exchanging two elements that are in inverse order reduces the number of inversions by at least one. Hence if there are i inversions, then with probability $\frac{1}{e}i\binom{n}{2}^{-1}$, the $(1+1)$ EA inverts one of the inversions and thus improves the fitness (the factor $\frac{1}{e}$ stems from only regarding iterations with $k = 0$, that is, where exactly one exchange is performed). By our fitness level theorem, the runtime T is dominated by the independent sum $\sum_{i=1}^{\binom{n}{2}} \text{Geom}(\frac{i}{e\binom{n}{2}})$. Hence

$$E[T] \leq e\binom{n}{2}H_{\binom{n}{2}} \leq \frac{e}{2}n^2(1 + 2\ln n),$$

$$\Pr[T \geq (1+\delta)en^2\ln n] \leq \binom{n}{2}^{-\delta},$$

where again the statement on the expectation has appeared before in [20] (with slightly different constants stemming from the use of a slightly different algorithm) and the tail bound is new.

4   Beyond the Fitness Level Theorem

Above we showed that in all situations where the classic fitness level method can be applied we immediately obtain that the runtime is dominated by a suitable sum of independent geometric distributions. We now show that the domination-by-distribution argument is not restricted to such situations. As an example, we use another combinatorial problem from [20], the single-source shortest path problem.

In [20], the single-source shortest path problem in a connected undirected graph $G = (V, E)$ with edge weights $w : E \to \mathbb{N}$ and source vertex $s \in V$ was solved via a $(1+1)$ EA as follows. Individuals are arrays of pointers such that each vertex different from the source has a pointer to another vertex. If, for a vertex v, following the pointers gives a path from v to s, then the length (=sum of weights of its edges) of this path is the fitness of this vertex; otherwise the fitness of this vertex is infinite. The fitness of an individual is the vector of the fitnesses of all vertices different from the source. In the selection step, an offspring is accepted if and only if all vertices have an at least as good fitness as in the parent ([20] call this a multi-objective fitness function). Mutating an individual means choosing a number k from a Poisson distribution with parameter $\lambda = 1$ and then changing $k + 1$ pointers to random new targets.

The main analysis argument in [20] is that, due to the use of the multi-objective fitness, a vertex that is connected to the source via a shortest path (let us call such a vertex *optimized* in the following) remains optimized for the remaining run of the algorithm. Hence we can perform a structural fitness level argument over the number of optimized vertices. The probability to increase this number is at least $p := \frac{1}{e(n-1)(n-2)}$ because there is at least one non-optimized vertex v for which the next vertex u on a shortest path from v to s is already optimized. Hence with probability $\frac{1}{e}$ we have $k = 0$, with probability $\frac{1}{n-1}$ we choose v, and with probability $\frac{1}{n-2}$ we rewire its pointer to u. This gives an expected optimization time of at most $E[T] \le (n-1)/p = e(n-1)^2(n-2)$.

To obtain better bounds for certain graph classes, [20] define n_i to be the number of vertices for which the shortest path with fewest edges consists of i edges. With a fitness level argument similar to the one above, they argue that it takes an expected time of at most $en^2 H_{n_1}$ to have all n_1-type vertices optimally connected to the source. After this, an expected number of at most $en^2 H_{n_2}$ iteration suffices to connect all n_2-vertices to the source. Iterating this argument, they obtain a runtime estimate of $E[T] \le en^2 \sum_{i=1}^{n-1} H_{n_i} \le en^2 \sum_{i=1}^{n-1} (\ln(n_i)+1)$. This expression remains of order $\Theta(n^3)$ in the worst case, but becomes, e.g., $O(n^2 \ell \log(\frac{2n}{\ell}))$ when each vertex can be connected to the source via a shortest path having at most ℓ edges.

With some technical effort, this result was improved to $O(n^2 \max\{\log n, \ell\})$ in [16]. We now show that this improvement could have been obtained via domination arguments in a very natural way.

Consider some vertex v different from the source. Fix some shortest path P from v to s having at most ℓ edges. Let V_v be the set of (at most ℓ) vertices on P different from s. As before, in each iteration we have a probability of at least $p = \frac{1}{e(n-1)(n-2)}$ that a non-optimized vertex of V_v becomes optimized. Consequently, the time T_v to connect v to s via a shortest path is dominated by a sum of ℓ independent $\mathrm{Geom}(p)$ random variables. We conclude that there are random variables X_{ij}, $i \in [1..\ell]$, $j \in [1..n-1]$, such that (i) $X_{ij} \sim \mathrm{Geom}(p)$ for all $i \in [1..\ell]$ and $j \in [1..n-1]$, (ii) for all $j \in [1..n-1]$ the variables $X_{1j}, \ldots, X_{\ell j}$ are independent, and (iii) the runtime T is dominated by $\max\{Y_j \mid j \in [1..n-1]\}$, where $Y_j := \sum_{i=1}^{\ell} X_{ij}$ for all $j \in [1..n-1]$.

It remains to deduce from this domination statement a runtime bound. Let $\delta = \max\{\frac{4\ln(n-1)}{\ell-1}, \sqrt{\frac{4\ln(n-1)}{\ell-1}}\}$ and $j \in [1..n-1]$. Then, by (8), $\Pr[Y_j \ge (1+\delta)E[Y_j]] \le \exp(-\frac{1}{2}\frac{\delta^2}{1+\delta}(\ell-1)) \le \exp(-\frac{1}{4}\min\{\delta^2, \delta\}(\ell-1)) \le \exp(-\frac{1}{4}\frac{4\ln(n-1)}{\ell-1}(\ell-1)) = \frac{1}{n-1}$. For all $\varepsilon > 0$, again by (8), we compute

$$\Pr[Y_j \ge (1+\varepsilon)(1+\delta)E[Y_j]] \le \exp\left(-\frac{1}{2}\frac{(\delta+\varepsilon+\delta\varepsilon)^2}{(1+\delta)(1+\varepsilon)}(\ell-1)\right)$$

$$\le \exp\left(-\frac{1}{2}\frac{\delta^2(1+\varepsilon)^2}{(1+\delta)(1+\varepsilon)}(\ell-1)\right)$$

$$\le \exp\left(-\frac{1}{2}\frac{\delta^2}{1+\delta}(\ell-1)\right)^{1+\varepsilon} \le (n-1)^{-(1+\varepsilon)}.$$

Recall that the runtime T is dominated by $Y = \max\{Y_j \mid j \in [1..n-1]\}$, where we did not make any assumption on the correlation of the Y_j. In particular, they do not need to be independent. Let $T_0 = (1+\delta)E[Y_1] = (1+\delta)\frac{\ell}{p}$. Then $\Pr[Y \ge (1+\varepsilon)T_0] \le \sum_{j=1}^{n-1} \Pr[Y_j \ge (1+\varepsilon)T_0] \le (n-1)^{-\varepsilon}$ by the union bound.

Transforming this tail bound into an expectation via standard arguments, we obtain $E[Y] \leq \left(1 + \frac{1}{\ln(n-1)}\right) T_0$.

In summary, with $\delta = \max\{\frac{4\ln(n-1)}{\ell-1}, \sqrt{\frac{4\ln(n-1)}{\ell-1}}\}$, $p = \frac{1}{e(n-1)(n-2)}$, and $T_0 :=$ $(1+\delta)\frac{\ell}{p}$, we find that the EA proposed in [20] solves the single-source shortest path problem in graphs where all vertices are connected to the source via a shortest path of at most ℓ edges, in a time T satisfying

$$E[T] \leq \left(1 + \frac{1}{\ln(n-1)}\right) T_0,$$
$$\Pr[T \geq (1+\varepsilon)T_0] \leq (n-1)^{-\varepsilon} \text{ for all } \varepsilon \geq 0.$$

Hence our domination argument proved the same result as the analysis in [16]. In fact, our analysis gives a better leading constant. For $\ell \gg \log(n)$, for example, our T_0 is $(1+o(1))e\ell n^2$, whereas the upper bound on the expected runtime in [16] is $(1+o(1))8e\ell n^2$.

5   Structural Domination

So far we have used stochastic domination to compare runtime distributions. We now show that stochastic domination can be a very useful tool also to express structural properties of the optimization process. As an example, we give a short and elegant proof for the result of Witt [2] that compares the runtimes of mutation-based algorithms. The main reason why our proof is significantly shorter than the one of Witt is that we use the notion of stochastic domination also for the distance from the optimum.

To state this result, we need the notion of a (μ, p) *mutation-based algorithm* introduced in [21]. This class of algorithms is called only *mutation-based* in [21], but since (i) it does not include all adaptive algorithms using mutation only, e.g., those regarded in [4,22–25], (ii) it does not include all algorithms using a different mutation operator than standard-bit mutation, e.g., those in [26–29], and (iii) this notion collides with the notion of unary unbiased black-box complexity algorithms (see [30]), which with some justification could also be called the class of mutation-based algorithms, we feel that a notion making these restrictions precise is more appropriate.

The class of (μ, p) mutation-based algorithms comprises all algorithms which first generate a set of μ search points uniformly and independently at random from $\{0, 1\}^n$ and then repeat generating new search points from any of the previous ones via standard-bit mutation with probability p (that is, by flipping bits independently with probability p). This class includes all $(\mu + \lambda)$ and (μ, λ) EAs which only use standard-bit mutation with static mutation rate p.

Denote by $(1+1)$ EA$_\mu$ the following algorithm in this class. It first generates μ random search points. From these, it selects uniformly at random one with highest fitness and then continues from this search point as a $(1+1)$ EA, that is, repeatedly generates a new search point from the current one via standard-bit mutation with rate p and replaces the previous one by the new one if the new one is not worse (in terms of the fitness). This algorithm was called $(1+1)$ EA with BestOf(μ) initialization in [31].

For any algorithm $\mathcal{A}$ from the class of (μ, p) mutation-based algorithms and any fitness function $f : \{0,1\}^n \to \mathbb{R}$, let us denote by $T(\mathcal{A}, f)$ the runtime of the algorithm $\mathcal{A}$ on the fitness function f, that is, the number of the first individual generated that is an optimal solution. Usually, this will be μ plus the number of the iteration in which the optimum was generated. To cover also the case that one of the random initial individuals is optimal, let us assume that these initial individuals are generated sequentially.

In this language, Witt [2] shows the following remarkable result.

Theorem 6. *For any (μ, p) mutation-based algorithm $\mathcal{A}$ and any $f : \{0,1\}^n \to \mathbb{R}$ with unique global optimum, $T((1+1)\ EA_\mu, \textsc{OneMax}) \preceq T(\mathcal{A}, f)$.*

This result significantly extends results of a similar flavor in [6,21,32]. The importance of such types of results is that they allow to prove lower bounds for the performance of many algorithm on essentially arbitrary fitness functions by just regarding the performance of the $(1+1)$ EA$_\mu$ on OneMax.

Let us denote by $|x|_1$ the number of ones in the bit string $x \in \{0,1\}^n$. Using similar arguments as in [33, Sect. 5] and [32, Lemma 13], Witt [2] shows the following natural domination relation between offspring generated via standard-bit mutation.

Lemma 4. *Let $x, y \in \{0,1\}^n$. Let $p \in [0, \frac{1}{2}]$. Let x', y' be obtained from x, y via standard-bit mutation with rate p. If $|x|_1 \leq |y|_1$, then $|x'|_1 \preceq |y'|_1$.*

We are now ready to give our alternate proof for Theorem 6. While it is clearly shorter that the original one in [2], we also feel that it is more natural. In very simple words, it shows that $T(\mathcal{A}, f)$ dominates $T((1+1)\ EA_\mu, \textsc{OneMax})$ because the search points generated in the run of the $(1+1)$ EA$_\mu$ on OneMax always are at least as close to the optimum (in the domination sense) as in the run of $\mathcal{A}$ on f, and this follows from the previous lemma and induction.

Proof. As a first small technicality, let us assume that the $(1+1)$ EA$_\mu$ in iteration $\mu + 1$ does not choose a random optimal search point, but the last optimal search point. Since all the first μ individuals are generated independently, this modification does not change anything.

Since $\mathcal{A}$ treats bit-positions and bit-values in a symmetric fashion (it is unbiased in the sense of [30]), we may without loss of generality assume that the unique optimum of f is $(1, \ldots, 1)$.

Let $x^{(1)}, x^{(2)}, \ldots$ be the sequence of search points generated in a run of $\mathcal{A}$ on the fitness function f. Hence $x^{(1)}, \ldots, x^{(\mu)}$ are independently and uniformly distributed in $\{0,1\}^n$ and all subsequent search points are generated from suitably chosen previous ones via standard-bit mutation with rate p. Let $y^{(1)}, y^{(2)}, \ldots$ be the sequence of search points generated in a run of the $(1+1)$ EA$_\mu$ on the fitness function OneMax.

We show how to couple these random sequences of search points in a way that $|\tilde{x}^{(t)}|_1 \leq |\tilde{y}^{(t)}|_1$ for all $t \in \mathbb{N}$. We take as common probability space Ω simply the space that $(x^{(t)})_{t \in \mathbb{N}}$ is defined on and let $\tilde{x}^{(t)} = x^{(t)}$ for all $t \in \mathbb{N}$.

We define the $\tilde{y}^{(t)}$ inductively as follows. For $t \in [1..\mu]$, let $\tilde{y}^{(t)} = x^{(t)}$. Note that this trivially implies $|\tilde{x}^{(t)}|_1 \leq |\tilde{y}^{(t)}|_1$ for these search points. Let $t > \mu$ and assume that $|\tilde{x}^{(t')}|_1 \leq |\tilde{y}^{(t')}|_1$ for all $t' < t$. Let $s \in [1..t-1]$ be maximal such that $|\tilde{y}^{(s)}|_1$ is maximal among $|\tilde{y}^{(1)}|_1, \ldots, |\tilde{y}^{(t-1)}|_1$. Let $r \in [1..t-1]$ be such that $x^{(t)}$ was generated from $x^{(r)}$ in the run of $\mathcal{A}$ on f. By induction, we have $|x^{(r)}|_1 \leq |\tilde{y}^{(r)}|_1$. By choice of s we have $|\tilde{y}^{(r)}|_1 \leq |\tilde{y}^{(s)}|_1$. Consequently, we have $|x^{(r)}|_1 \leq |\tilde{y}^{(s)}|_1$. By Lemma 4 and Theorem 1, there is a random $\tilde{y}^{(t)}$ (defined on Ω) such that $\tilde{y}^{(t)}$ has the distribution of being obtained from $\tilde{y}^{(s)}$ via standard-bit mutation with rate p and such that $|x^{(t)}|_1 \leq |\tilde{y}^{(t)}|_1$.

With this construction, the sequence $(\tilde{y}^{(t)})_{t \in \mathbb{N}}$ has the same distribution as $(y^{(t)})_{t \in \mathbb{N}}$. This is because the first μ elements are random and then each subsequent one is generated via standard-bit mutation from the current-best one, which is just the way the $(1+1)$ EA$_\mu$ is defined. At the same time, we have $|\tilde{x}^{(t)}|_1 \leq |\tilde{y}^{(t)}|_1$ for all $t \in \mathbb{N}$. Consequently, we have $\min\{t \in \mathbb{N} \mid |\tilde{y}^{(t)}|_1 = n\} \leq \min\{t \in \mathbb{N} \mid |x^{(t)}|_1 = n\}$. Since $T((1+1)\,\text{EA}_\mu, \text{OneMax})$ and $\min\{t \in \mathbb{N} \mid |\tilde{y}^{(t)}|_1 = n\}$ are identically distributed and also $T(\mathcal{A}, f)$ and $\min\{t \in \mathbb{N} \mid |x^{(t)}|_1 = n\}$ are identically distributed, we have $T((1+1)\,\text{EA}_\mu, \text{OneMax}) \preceq T(\mathcal{A}, f)$. $\square$

6   Conclusion

In this work, we argued that stochastic domination can be very useful in runtime analysis, both to formulate more informative results and to obtain simpler and more natural proofs. We also showed that in many situations, in particular, whenever the fitness level method is applicable, it is easily possible to describe the runtime via a domination statement.

We note however that not all classic proofs easily reveal details on the distribution. For results obtained via random walk arguments, e.g., the optimization of the short path function SPC$_n$ [34], monotone polynomials [35], or vertex covers on paths-like graphs [36], as well as for results proven via additive drift [37], the proofs often give little information about the runtime distribution (an exception is the analysis of the needle and the OneMax function in [38]).

For results obtained via the average weight decrease method [39] or multiplicative drift analysis [18], the proofs also do not give information on the runtime distribution. However, the probabilistic runtime bound of type $\Pr[T \geq T_0 + \lambda] \leq (1 - \delta)^\lambda$ obtained from these methods implies that the runtime is dominated by $T \preceq T_0 - 1 + \text{Geom}(1 - \delta)$.

Overall, both from regarding these results and the history of the field, we suggest to more frequently formulate results via domination statements. Even in those cases where the probabilistic tools at the moment are not be ready to exploit such a statement, there is a good chance future developments overcome this shortage and then it pays off if the result is readily available in a distribution form and not just as an expectation.

References

1. Droste, S., Jansen, T., Wegener, I.: On the analysis of the (1+1) evolutionary algorithm. Theoret. Comput. Sci. **276**, 51–81 (2002)
2. Witt, C.: Tight bounds on the optimization time of a randomized search heuristic on linear functions. Comb. Probab. Comput. **22**, 294–318 (2013)
3. Doerr, B., Jansen, T., Witt, C., Zarges, C.: A method to derive fixed budget results from expected optimisation times. In: Genetic and Evolutionary Computation Conference, GECCO 2013, pp. 1581–1588. ACM (2013)
4. Böttcher, S., Doerr, B., Neumann, F.: Optimal fixed and adaptive mutation rates for the LeadingOnes problem. In: Schaefer, R., Cotta, C., Kołodziej, J., Rudolph, G. (eds.) PPSN 2010. LNCS, vol. 6238, pp. 1–10. Springer, Heidelberg (2010). https://doi.org/10.1007/978-3-642-15844-5_1
5. Jansen, T., Zarges, C.: Performance analysis of randomised search heuristics operating with a fixed budget. Theoret. Comput. Sci. **545**, 39–58 (2014)
6. Borisovsky, P.A., Eremeev, A.V.: Comparing evolutionary algorithms to the (1+1)-EA. Theoret. Comput. Sci. **403**, 33–41 (2008)
7. Zhou, D., Luo, D., Lu, R., Han, Z.: The use of tail inequalities on the probable computational time of randomized search heuristics. Theoret. Comput. Sci. **436**, 106–117 (2012)
8. Witt, C.: Fitness levels with tail bounds for the analysis of randomized search heuristics. Inf. Process. Lett. **114**, 38–41 (2014)
9. Doerr, B., Doerr, C.: A tight runtime analysis of the (1+(λ, λ)) genetic algorithm on OneMax. In: Proceedings of the Genetic and Evolutionary Computation Conference, GECCO 2015, pp. 1423–1430. ACM (2015)
10. Müller, A., Stoyan, D.: Comparison Methods for Stochastic Models and Risks. Wiley, Hoboken (2002)
11. Doerr, B., Happ, E., Klein, C.: Crossover can provably be useful in evolutionary computation. Theoret. Comput. Sci. **425**, 17–33 (2012)
12. Doerr, B.: Analyzing randomized search heuristics: tools from probability theory. In: Auger, A., Doerr, B. (eds.) Theory of Randomized Search Heuristics, pp. 1–20. World Scientific, Singapore (2011)
13. Wegener, I.: Theoretical aspects of evolutionary algorithms. In: Orejas, F., Spirakis, P.G., van Leeuwen, J. (eds.) ICALP 2001. LNCS, vol. 2076, pp. 64–78. Springer, Heidelberg (2001). https://doi.org/10.1007/3-540-48224-5_6
14. Janson, S.: Tail bounds for sums of geometric and exponential variables. arXiv e-prints arXiv:1709.08157 (2017)
15. Scheideler, C.: Probabilistic methods for coordination problems. University of Paderborn, Habilitation thesis (2000). http://citeseerx.ist.psu.edu/viewdoc/summary?doi=10.1.1.70.1319
16. Doerr, B., Happ, E., Klein, C.: Tight analysis of the (1+1)-EA for the single source shortest path problem. Evol. Comput. **19**, 673–691 (2011)
17. Doerr, B., Doerr, C.: The impact of random initialization on the runtime of randomized search heuristics. Algorithmica **75**, 529–553 (2016)
18. Doerr, B., Goldberg, L.A.: Adaptive drift analysis. Algorithmica **65**, 224–250 (2013)
19. Doerr, B., Künnemann, M.: Optimizing linear functions with the (1+λ) evolutionary algorithm–different asymptotic runtimes for different instances. Theoret. Comput. Sci. **561**, 3–23 (2015)

20. Scharnow, J., Tinnefeld, K., Wegener, I.: The analysis of evolutionary algorithms on sorting and shortest paths problems. J. Math. Model. Algorithms **3**, 349–366 (2004)
21. Sudholt, D.: A new method for lower bounds on the running time of evolutionary algorithms. IEEE Trans. Evol. Comput. **17**, 418–435 (2013)
22. Jansen, T., Wegener, I.: On the analysis of a dynamic evolutionary algorithm. J. Discrete Algorithms **4**, 181–199 (2006)
23. Oliveto, P.S., Lehre, P.K., Neumann, F.: Theoretical analysis of rank-based mutation - combining exploration and exploitation. In: Congress on Evolutionary Computation, CEC 2009, pp. 1455–1462. IEEE (2009)
24. Badkobeh, G., Lehre, P.K., Sudholt, D.: Unbiased black-box complexity of parallel search. In: Bartz-Beielstein, T., Branke, J., Filipič, B., Smith, J. (eds.) PPSN 2014. LNCS, vol. 8672, pp. 892–901. Springer, Cham (2014). https://doi.org/10.1007/978-3-319-10762-2_88
25. Doerr, B., Gießen, C., Witt, C., Yang, J.: The $(1+\lambda)$ evolutionary algorithm with self-adjusting mutation rate. In: Genetic and Evolutionary Computation Conference, GECCO 2017. ACM (2017)
26. Doerr, B., Doerr, C., Yang, J.: k-bit mutation with self-adjusting k outperforms standard bit mutation. In: Handl, J., Hart, E., Lewis, P.R., López-Ibáñez, M., Ochoa, G., Paechter, B. (eds.) PPSN 2016. LNCS, vol. 9921, pp. 824–834. Springer, Cham (2016). https://doi.org/10.1007/978-3-319-45823-6_77
27. Doerr, B., Doerr, C., Yang, J.: Optimal parameter choices via precise black-box analysis. In: Genetic and Evolutionary Computation Conference, GECCO 2016, pp. 1123–1130. ACM (2016)
28. Lissovoi, A., Oliveto, P.S., Warwicker, J.A.: On the runtime analysis of generalised selection hyper-heuristics for pseudo-boolean optimisation. In: Genetic and Evolutionary Computation Conference, GECCO 2017, pp. 849–856. ACM (2017)
29. Doerr, B., Le, H.P., Makhmara, R., Nguyen, T.D.: Fast genetic algorithms. In: Genetic and Evolutionary Computation Conference, GECCO 2017. ACM (2017)
30. Lehre, P.K., Witt, C.: Black-box search by unbiased variation. Algorithmica **64**, 623–642 (2012)
31. de Perthuis de Laillevault, A., Doerr, B., Doerr, C.: Money for nothing: speeding up evolutionary algorithms through better initialization. In: Genetic and Evolutionary Computation Conference, GECCO 2015, pp. 815–822. ACM (2015)
32. Doerr, B., Johannsen, D., Winzen, C.: Multiplicative drift analysis. Algorithmica **64**, 673–697 (2012)
33. Droste, S., Jansen, T., Wegener, I.: A natural and simple function which is hard for all evolutionary algorithms. In: IEEE International Conference on Industrial Electronics, Control, and Instrumentation, IECON 2000, pp. 2704–2709. IEEE (2000)
34. Jansen, T., Wegener, I.: Evolutionary algorithms - how to cope with plateaus of constant fitness and when to reject strings of the same fitness. IEEE Trans. Evol. Comput. **5**, 589–599 (2001)
35. Wegener, I., Witt, C.: On the optimization of monotone polynomials by simple randomized search heuristics. Comb. Probab. Comput. **14**, 225–247 (2005)
36. Oliveto, P.S., He, J., Yao, X.: Analysis of the $(1+1)$-EA for finding approximate solutions to vertex cover problems. IEEE Trans. Evol. Comput. **13**, 1006–1029 (2009)

37. He, J., Yao, X.: Drift analysis and average time complexity of evolutionary algorithms. Artif. Intell. **127**, 51–81 (2001)
38. Garnier, J., Kallel, L., Schoenauer, M.: Rigorous hitting times for binary mutations. Evol. Comput. **7**, 173–203 (1999)
39. Neumann, F., Wegener, I.: Randomized local search, evolutionary algorithms, and the minimum spanning tree problem. Theoret. Comput. Sci. **378**, 32–40 (2007)

On the Fractal Nature of Local Optima Networks

Sarah L. Thomson[1]($\boxtimes$)(iD), Sébastien Verel[2](iD), Gabriela Ochoa[1](iD),
Nadarajen Veerapen[1](iD), and Paul McMenemy[1](iD)

[1] Computing Science and Mathematics, University of Stirling, Stirling, UK
{s.l.thomson,gabriela.ochoa,nadarajen.veerapen,paul.mcmenemy}@stir.ac.uk
[2] Université du Littoral Côte d'Opale, EA 4491 - LISIC, Calais, France
verel@uni-littoral.fr

Abstract. A *Local Optima Network* represents fitness landscape connectivity within the space of local optima as a mathematical graph. In certain other complex networks or graphs there have been recent observations made about inherent *self-similarity*. An object is said to be self-similar if it shows the same patterns when measured at different scales; another word used to convey self-similarity is *fractal*. The *fractal dimension* of an object captures how the detail observed changes with the scale at which it is measured, with a high fractal dimension being associated with complexity. We conduct a detailed study on the fractal nature of the local optima networks of a benchmark combinatorial optimisation problem (NK Landscapes). The results draw connections between fractal characteristics and performance by three prominent metaheuristics: Iterated Local Search, Simulated Annealing, and Tabu Search.

Keywords: Combinatorial fitness landscapes
Local optima networks · Fractal analysis · NK Landscapes

1 Introduction

Weinberger and Stadler [1] noticed that certain fitness landscapes exhibit *self-similarity*. They saw that if they increased landscape diameter, they observed patterns of ruggedness that scaled in a way indicative of fractal geometry. In particular, the landscapes showed evidence of having a multilevel structure. The critical question is then whether we can exploit the fractal patterns.

The information used in the aforementioned study was obtained by conducting random walks on the fitness landscape at the solution level. Considering the solution level to be the base of the search space, we can then consider fractal patterns at higher levels of abstraction; to see the extent of fractal geometry in a model of the local optima space would be desirable.

The original version of this chapter was revised: Figure 4 was corrected. The erratum to this chapter is available at https://doi.org/10.1007/978-3-319-77449-7_13

© Springer International Publishing AG, part of Springer Nature 2018
A. Liefooghe and M. López-Ibáñez (Eds.): EvoCOP 2018, LNCS 10782, pp. 18–33, 2018.
https://doi.org/10.1007/978-3-319-77449-7_2

A *local optima network* [2] models the local optima level of a fitness landscape. A network is formed by tracing the search connectivity between optima where the network nodes are local optima. An edge traced between two nodes means that the destination (optimum) can be reached from the source (optimum) by carrying out a perturbation followed by hill-climbing. This measure of distance captures the notion of neighbourhood in the space of local optima.

Conducting a fractal analysis of a local optima network would give information about patterns at a raised level of abstraction in the fitness landscape. Of particular interest is the *fractal dimension* [3], hereafter denoted as FD. Here, a non-integer dimension can be assigned to a shape as a measure of complexity. This creates an index which quantifies how detail in an object is observed when the scale at which it is measured is changed. Figure 1a and b provide examples of this; both are shapes with FD somewhere between one and two, but they have markedly different complexities. We can see that the latter displays a significantly more complex pattern composition, containing much more detail at different scales than the former and filling much more of the overall space which it occupies. The shape in Fig. 1a contains a lot of empty space—a lack of 'space-filling' nature—and therefore low fractional dimension. Another perspective is to consider how different from the one-dimensional line the two are. Figure 1a is nearer to one dimension than two, at 1.1292; indeed, we can see that it is effectively a line with some ruggedness or detail. Figure 1b is almost two-dimensional, but is missing complicated segments.

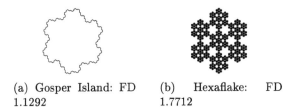

(a) Gosper Island: FD 1.1292   (b)   Hexaflake: FD 1.7712

Fig. 1. Two fractals with different dimensions.

The study of complex networks as a field in its own right has created a wealth of measures to understand them, including algorithms for calculating the *FD* of a complex network. One of these, termed a 'box-counting' algorithm [4], tries to describe a network with as few *boxes* as possible, with each box containing nodes which are within m links of each other. The parameter m corresponds to the length of measure used in the equation to obtain fractal dimension. The number of boxes (as a proportion of the size of the network) is taken to be the extent of detail observed in the shape when using the scale m.

Because this method is agnostic of the semantics of the network, it can also be used to calculate the FD of a local optima network. However, distance between nodes is not the only important consideration. Node fitness, as well as link-distance, is of great significance in a local optima network. Accordingly, a modification of the box-count algorithm is required. A threshold should be incorporated, for the maximum fitness difference allowed between nodes which can be boxed together. In this way, nodes which satisfy both a distance and fitness criterion can be grouped, and the FD calculated from the end result.

This study is intended as an introductory investigation into the use of fractal measures in the space of local optima in a fitness landscape. We compute the FD and associated metrics for a set of NK Landscape instances. The obtained results suggest links between the fractal geometry in the networks and the empirical difficulty for search algorithms.

2   Background

2.1   The Study of Fitness Landscapes

A fitness landscape [5] is a triplet (S, N, f) where S is the set of all possible solutions, $N : S \longrightarrow 2^S$, a neighbourhood structure, is a function that assigns to every $s \in S$ a set of neighbours $N(s)$, and f is a fitness (objective value) function such that $f : S \longrightarrow \mathbb{R}$, where the fitness value is a real number that can be viewed as the *height* of a given solution in the landscape.

2.2   The Local Optima Network

A *local optima network* is a representation of the fitness landscape at the level of local optima. We now formally define the constituent parts of a local optima network, before proceeding to describe the object as a whole.

Nodes. The set of nodes, LO, is comprised of local optima, i.e. a solution lo_i satisfies the condition that it has superior fitness to all other solutions in its neighbourhood: $\forall n \in N(lo_i) : f(lo_i) \geq f(n)$, where $N(lo_i)$ is the neighbourhood and n is a single neighbour.

Edges. The set of edges, E, consists of directed and weighted links. An edge is traced if the probability of 'escape'—using perturbation and then hill-climbing—from the source node to the destination is greater than zero, and is weighted with the probability. Formally, local optima lo_i and lo_j form the source and destination of an edge iff $w_{ij} > 0$.

Local optima network (LON). The weighted local optima network $LON = (LO, E)$ is a graph where the nodes $lo_i \in LO$ are the local optima, and there exists an edge $e_{ij} \in E$, with weight w_{ij}, between two nodes lo_i and lo_j if $w_{ij} > 0$. Note that w_{ij} may be different than w_{ji}. Thus, two weights are needed in general, and so a local optima network is an oriented transition graph.

2.3   The Fractal Dimension

This study aims to draw links between the fractal detail in local optima networks and search success on the underlying problem instances. Specifically, the *fractal dimension* [3] (FD) is used to characterise the dimensional complexity of the networks. To understand FD we start with the more familiar notion of typical geometric dimensions; the one-dimensional line, for example, or the two-dimensional square. A line is one-dimensional, as if we try to measure it with a scale half the length of the line (we zoom in by a factor of two), we measure or obtain exactly two copies of the original shape. In general, to calculate dimension for a given object, we need the scaling factor (two in this example), and the extent of detail observed using that scale (here, the two copies). The result is found by solving for x the equation

$$scale^x = detail \tag{1}$$

which in this example is $2^x = 2$. The value for x is one, meaning it is a one-dimensional object.

It follows that a shape containing fractal geometry does not have an integer dimension. Instead it has a fractal dimension, which lies somewhere on the real number line. An object with dimension just above an integer (for example 2.12) is only slightly more complex or detailed than the dimension below. Conversely, a dimension just below an integer (e.g. 2.89) indicates a shape with that dimension but with complicated patterns removed. In essence, fractal dimension is a complexity index and captures how detail in an object relates to a scaling factor. Another way to approach this notion is to consider how well a pattern fills the geometric space where it resides. For example, recall Fig. 1b, where the patterns fill much of the possible space. This shape has a high FD; the way it fills space is much more complicated than a shape with a smaller integer dimension. Rearranging Eq. 1, we obtain the fractal dimension:

$$FD = \frac{\log(detail)}{\log(scale)} \tag{2}$$

2.4   Fractals and Fitness Landscapes

Weinberger and Stadler [1] noticed self-similar behaviour in certain fitness landscapes. They used the well-known *autocorrelation* metric [6] in their analysis. The way autocorrelation scaled with landscape diameter was indicative, in some cases, of fractal geometry.

Several years later, Locatelli conducted a detailed study [7] on the phenomenon of patterns re-appearing at different levels of abstraction in fitness landscapes. They termed this the 'multilevel' structure of optimisation problems, and noted that it could be exploited.

Zelinka *et al.* [8] also demonstrated the potential of using fractal analysis for learning more about the nature of fitness landscapes, focusing on low-dimensional, continuous spaces.

Until now, there has been a lack of study regarding fractal patterns within a local optima network. One consideration is how precisely to define the dimensionality: a complex network is quite different to the typical two-dimensional pictures used in fractal analysis. Methods have been proposed for calculating dimension in the specific case of a network. In this study, a 'box-counting' algorithm is used and extended to cater for the local optima network case.

2.5   Fractals and Complex Networks

We use a state-of-the-art 'box-counting' methodology to define the FD of a complex network, proposed in Song et al. [4]. The process iteratively boxes together nodes which are within m links of each other, with the aim of describing the network in as few boxes as possible. The parameter m is an integral part of computing the FD: it is one of the two values needed for the relation between detail and scale, corresponding to the *scale* parameter introduced in Sect. 3. For *detail*, we calculate the number of boxes (when the algorithm has converged) as a proportion of total network size, P.

In the specific case of a *local optima* network, link-distance is not the only consideration when boxing together nodes. Crucial information about landscape structure is encoded in the fitness values of the local optima. The process of boxing identifies how much *detail* is observed measuring the object at a certain scale. Omitting the fitness values would ignore fitness detail (for example, a large fitness difference between two optima near each other in the space). Therefore, an extension of the algorithm is required. Two nodes should be boxed together only if they satisfy both a distance *and* a fitness condition. More specifically, two nodes, lo_i and lo_j, can be boxed together in the algorithm if the distance between them is less than m, where m is the maximum links allowed between nodes boxed together, and $|f(lo_i) - f(lo_j)| < \epsilon$, where ϵ is the maximum allowed fitness disparity. In subsequent text, we denote the link distance between two nodes in the network as $d(lo_i, lo_j)$.

Pseudo-code for iterative box-counting of a local optima network is shown in Algorithm 1. The notation $mass(v)$ is used to represent the 'mass' of the network which can be covered using the vertex v as a reference point. Upon convergence of the algorithm, we need to examine the relation between detail and scaling to obtain the FD. We have the number of boxes needed, b, which we can take as a proportion of the network as P. For the scaling factor, we use the link-distance parameter m. This corresponds to the level of abstraction being applied to measure the shape. We insert P and m into Eq. 2 to derive the FD of the local optima network.

3   Experimental Setting

3.1   Test Problem

We consider instances from a benchmark combinatorial optimisation domain in this work: the NK Landscape model. The instances are from the work by Ochoa et al. [2]. The problems are deliberately small in size, such that a full enumeration of the local optima is possible. This is particularly necessary due to the introductory nature of this study into whether fractal analysis of local optima networks is helpful.

NK Landscapes are a family of synthetic fitness functions. They give rise to fitness landscapes which can be tuned from completely smooth to completely

---

Algorithm 1. Box-counting a Local Optima Network

Initialisation:
V, CV, NCV = nodes in network, center nodes, non-center nodes
$CV = [\,]$, $NCV = V$
$Cov, NCov$ = covered nodes, non covered nodes
$NCov = V, Cov = [\,]$
Stage 1:
repeat
    for v in NCV **do**
        mass(v) = count(v' in NCov where $d(v, v') < mb$ and diff($f_v, f_{v'}) < \epsilon$)
    end for
    $next.center = v$ where $mass(v) == max(mass(v \in NCV))$
    for v in NCov **do**
        $distance =$ d($next.center, v$)
        $e =$ diff($f(next.center), f(v)$)
        if $e < \epsilon$ and $distance < mB$ **then**:
            $NCov = NCov - v$
        end if
    end for
until $\forall v \in V, v \in Cov$ or $v \in CV$
Stage 2:
for c in CV **do**
    $id(c) =$ generate(distinct id)
    for v in V **do**
        lowest = lowest combined fitness-distance and link-distance seen so far
        closest = centre node which is closest
        for c in CV **do**
            $distance =$ d(v,c), e = diff(f_v, f_c)
            if $distance + e < lowest$ and $distance < mB$ **then**:
                $closest = c$
                $lowest = distance + e$
            end if
        end for
        $closest.centers[v] =$ lowest
    end for
end for
$NCov = NCov$ according to $closest.centers$ values (ascending)
for for v in NCV **do**
    $v' =$ neighbour of v with lower value in $closest.centers$ array
    $id(v) = id(v')$
    remove v from NCV
end for

---

random. There are two parameters: N and K. Solutions are binary-encoded and of length N. The parameter K dictates how many of the binary variables are dependent on each other—epistasis. Each bit has a numeric value assigned from a uniform distribution of floating-point numbers. The fitness of a given solution, S, is the average of the fitness contributions of the N bits:

$$F(S) = \sum^i f(s_i) \tag{3}$$

when calculating the contribution of a bit, s_i, the values of K other bits are considered:

$$f(s(i)) = f(s_i, s_1^i, ..., s_k^i) \tag{4}$$

Here we use $N = 18$, and $K = \{2,4,6,8\}$, with 30 instances for each K, for a total of 120. NK Landscapes are often used as a test-bed for new fitness landscape techniques because ruggedness can be introduced in a controlled way, by increasing the value for K.

3.2   Metaheuristics

For the fractal analysis to be useful, a view of the relationship between FD and search difficulty in the underlying problem instances should be sought. However, the notion of difficulty is subjective to the algorithm used. We deploy common trajectory-based metaheuristics: iterated local search (ILS), simulated annealing (SA) and tabu search (TS). For the local search element of all three algorithms, a bit-flip is deployed. For the perturbation in the case of the ILS, this mechanism is doubled. The SA parameters are those suggested in Thomson *et al.* [9] in a study on NK Landscapes. The start and end temperatures are 1.4 and 0.0, respectively; α is set at 0.8; and the maximum iterations at the same temperature is 262. The length of the tail in the tabu search is set at n, i.e. 18, the length of the solutions.

All three metaheuristics were implemented with *Paradiseo* [10], an open-source package in C++, and were run 1000 times each per problem instance.

3.3   Fractal Analysis

To calculate the FD of our networks, we employ Algorithm 1: a box-counting algorithm specialised to the local optima network case. An implementation for box-counting a network in C was obtained from the work in Song *et al.* [4].

As we stated in Sect. 2.5, two important parameters exist in our consideration of FD in the specific case of a local optima network. The parameter m, used in the general box-count algorithm, controls edge-distance between boxed nodes; we also have the parameter ϵ, which is the maximum fitness difference allowed between boxed nodes. This should operate on normalised fitness ranges, such that a single ϵ value can be applied to all networks. A fitness value can be normalised as $f = \frac{f - E(f)}{sd(f)}$, where $E(f)$ is the expected fitness value and $sd(f)$ is the standard deviation. With this, the mean becomes zero while the standard deviation is one.

4   Results

We compute the FD of the local optima networks for each of the 120 problem instances considered. This is done using the modified box-counting algorithm outlined in Sect. 3.3, and represented in Algorithm 1. For each network, we compute 20 FDs, differentiated by setting the fitness-mandate parameter ϵ, which is in the range $[0.0, 1.0]$ in step sizes of 0.05.

The three metaheuristics described in Sect. 3.2 are applied to each problem instance. Then, with the fractal information, the feature information and the performance data from the metaheuristics we can proceed to examine the relationships between these measures. The ultimate aim of the experiments in this study was to assess whether fractal analysis of a local optima network could be helpful.

4.1   Fractals and Epistasis

Firstly, we examine the distribution of FD in the local optima networks. The instances can be split by their *epistasis* level, i.e. how many variables are interdependent in a solution. Figure 2 shows box-plots for each of the four levels used.

We can see from the Fig. 2 that the FDs of the local optima networks increase with the epistatic parameter K, which is known to increase ruggedness and randomness in NK Landscapes. We can see this by, for example, comparing the interquartile range in category $K2$ with category $K4$. This would suggest that more ordered, predictable problem instances give rise to lower dimension objects. The interquartile range for the $K = 2$ instances spans ~1.008 to ~1.379. The median of the K2 LON dimension is 1.260. These are objects between one and two-dimensional; somewhere between a line and a 2-d shape. The dimension is closer to one, implying a structure akin to a 2d shape but with lots of parts removed from it, bringing it closer to resembling a line. In our context, the shapes we are dealing with are local optima networks, which represent search connectivity between local optima. A dimension such as this might imply these networks form somewhat linear sequences, with some deviation.

The instances with $K = 4$ generally have higher fractional dimension, with the interquartile range spanning ~1.363 to ~1.878. Again this implies that the networks are located between one and two dimensions. This time they are closer to two, with a median of 1.623. The fractal in Fig. 3 has dimension ~1.585. We can view it as a 2-d triangle which has complicated portions cut out of it. Of course, the structure of a local optima network with this dimension is unlikely to look like this; however, the way the detail in local optima network scales is similar to this fractal. Having a dimension of around 1.585 implies that when we reduce our scale to measure the shape in Fig. 3 to one fourth (a scaling factor of four), we obtain *nine* copies of our original shape. Using Eq. 1, i.e. solving $4^x = 9$ for x, does not result in an integer dimension, but instead ~1.585, a fractional dimension.

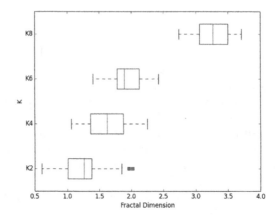

Fig. 2. Boxplots showing the FD of the local optima networks, grouped by epistasis level; 30 instances of each group are considered for each of $K = \{2,4,6,8\}$, all with $N = 18$.

Fig. 3. A shape with dimension between one and two. Sierpinski triangle—FD ~1.585.

Consider again Fig. 2, this time for the $K = 6$ instances. Here the interquartile range spans ~1.779 to ~2.126, with a median value of ~1.893. This tells us that the networks of local optima are either highly complex in being just below two-dimensional (see Fig. 1b for a fractal like this), or three-dimensional with a lot of the shape absent, leaning towards a 2d shape.

Notice that the increase in dimension is stark between the K6 and K8 groups. While the majority of dimensions in the lower-epistasis LONs were somewhere between one and two, these highly rugged fitness landscapes appear to give rise to local optima networks with dimension mostly between three and four. Of course, this situation is virtually impossible to visualise or conceptualise.

4.2 Fractal Dimension and Search Performance

We now consider how the FD of local optima networks is connected to empirical search difficulty. To gain insight into this, we must select appropriate measures of hardness.

For search algorithms, two things matter: efficiency and effectiveness. To quantify these, we should have a measure of success (or lack thereof) and a measure of speed. Here we use a proportional measure to represent success: the number of runs which reached the global optimum divided by the total runs. To measure efficiency, we consider the number of function evaluations used in successful runs.

Correlation Study. A logical place to start when contrasting problem features with performance is by correlation study. We compute the Spearman correlation coefficient and corresponding p-value for the observed fractal features and the performance measures.

In the interest of space, Fig. 4 shows the correlation matrix for a selected sample of features of the 120 considered problems. We show the correlation coefficient between variables (upper triangle of plot) density plots (middle diagonal), and scatter-plots (lower triangle). In the case of the density and scatter-plots, difference in colour indicates a split into epistasis level, as outlined in the caption. The size of the text is proportional to the strength of the absolute value of the correlation. An indication of p-value level is given by the asterisk, as described in the caption.

We include all measures of metaheuristic performance in the variable set. There are two metrics for each algorithm, giving six in total; one each for efficiency, and one for effectiveness. Indication of the measure is given after the abbreviation for the algorithm name; for example, *ILS.s* is the success rate of the ILS, while *ILS.t* is the run-time.

The remaining variables included are fitness landscape features. We are particularly interested in those relating to fractal geometry in the local optima networks. In our experiments, we computed correlations between performance and dimension at four different values for ϵ, namely $\{0.25, 0.50, 0.75, 1.00\}$, operating on normalised fitness ranges. The lower the value of this parameter, the more strict the fractal analysis algorithm is when boxing together nodes: recall that they are considered neighbours iff $d(lo_i, lo_j) < d$ and $|f(lo_i) - f(lo_j)| < \epsilon$. If ϵ is nearer to 1.0, the condition for grouped nodes is more lenient.

In the Figure, we show FD correlations with this parameter set at 0.5 (*FD1*) and 1.0 (*FD2*), but 0.25 and 0.75 showed similar trends.

In addition to the fractal dimension relationships, we show those of the number of local optima (*num.optima*) and of the number of landscape funnels (*num.funnels*). A *funnel* is a basin of attraction at the level of local optima and has been linked to search complications in NK Landscape problems [9].

Surveying the correlation matrix, the drastic effect of the parameter ϵ—the local optima network-specific detail parameter—on the resultant FD can be seen as quite apparent when comparing the two FD columns. We can check against the intersections with the performance metric rows. When set at 0.50 (*FD1*), dimension is positively associated with success rate for all three algorithms. In all cases, there appears to be statistical validity. Looking at the runtime rows, we can see that there is a positive connection with SA runtime, but negative ones with ILS and TS. All relationships between this particular FD and performance have **$p < 0.01$.

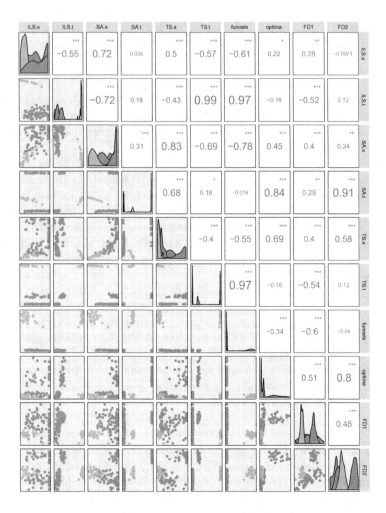

Fig. 4. Correlation matrices of performance metrics and landscape features (see facet titles). Lower triangle: pairwise scatter plots. Diagonal: density plots. Upper triangle: pairwise Spearman's rank correlation, $^{***}p < 0.001$, $^{**}p < 0.01$, $^{*}p < 0.05$. Colour represents instances split into different levels of epistasis ($K \in \{2, 4, 6, 8\}$). (Color figure online)

If we now consider the other FD, where ϵ is set leniently at 1.0 (column *FD2*), there are quite different results. For the two algorithms based on single-strength mutation (SA; TS), there appears to be a statistically significant connection with success rate. For all algorithms, the correlation with runtime is positive; however, of these, only that of SA is significant, and indeed this is one of the strongest relationships on the matrix.

Regression Models. Using linear models which include the FD as a predictor and a performance metric as the dependent variable allows us to further examine connections with search difficulty. We can normalise the predictors used as $p = \frac{p-E(p)}{sd(p)}$, where $E(p)$ is the expected predictor value and $sd(p)$ is the standard deviation. In this way, we can examine the strengths of contribution from problem features and gain a view of how much variance seen in the response variable is attributable to those we are interested in—the fractal measures.

Table 1. Predictor variables used in the regression models.

Notation	Description
Optima	Number of local optima
Funnels	Number of basins of attraction occuring in the space of local optima
Fractal[1]	FD with fitness difference threshold ϵ set at 0.25 (strictest)
Fractal[2]	FD with fitness difference threshold ϵ set at 0.50
Fractal[3]	FD with fitness difference threshold ϵ set at 0.75
Fractal[4]	FD with fitness difference threshold ϵ set at 1.00 (most lenient)

Table 2. Linear mixed models. The dependent variables are respective metaheuristic success rates. $^{***}p < 0.001$, $^{**}p < 0.01$, $^{*}p < 0.05$.

Predictor	ILS Estimate	SA Estimate	TS Estimate
Fractal Dim[1]	**0.605**(0.271)*	**0.483**(0.169)**	0.298(0.202)
Fractal Dim[2]	−0.313(0.202)	**−0.531**(0.126)***	**0.575**(0.151)***
Fractal Dim[3]	−0.098(0.145)	−0.122(0.090)	0.164(0.108)
Fractal Dim[4]	0.023(0.083)	**0.174**(0.052)***	**0.306**(0.062)***
Optima	0.000(0.000)	0.000(0.000)	0.000(0.000)*
Funnels	−0.003(0.001)***	−0.004(0.000)***	−0.004(0.001)***
R^2	0.380	0.709	0.685

Table 2 summarises three linear models: one for the success rate of each metaheuristic considered. The predictors used are introduced in Table 1.

The model summaries include coefficient estimates, standard errors (shown in parentheses), indication of p-value (see caption), and the adjusted R^2 value for the model.

Noteable results are shown in boldened text. Looking at the R^2 values, we can see that the ILS model is considerably weaker than those of the SA and TS.

Table 3. Linear mixed models. The dependent variables are respective metaheuristic run-times. $^{***}p < 0.001$, $^{**}p < 0.01$, $^*p < 0.05$.

Predictor	ILS Estimate	SA Estimate	TS Estimate
Fractal Dim^1	$-0.183(0.481)$	$-0.281(0.549)$	$-0.409(0.504)$
Fractal Dim^2	$-0.228(0.358)$	$\mathbf{-1.994}(0.409)^{***}$	$\mathbf{-0.854}(0.376)^*$
Fractal Dim^3	$-0.025(0.257)$	$-0.055(0.293)$	$-0.133(0.269)$
Fractal Dim^4	$\mathbf{0.402}(0.147)^{**}$	$\mathbf{1.512}(0.168)^{***}$	$\mathbf{0.538}(0.154)^{***}$
Optima	$0.000(0.000)^{**}$	$-0.001(0.000)^{***}$	$0.001(0.000)^{***}$
Funnels	$0.049(0.001)^{***}$	$0.000(0.001)$	$0.056(0.001)^{***}$
R^2	0.9743	0.918	0.979

Nonetheless, there is one fractal metric which appears to have statistical significance as a predictor in this model: Fractal Dim^1, which is the FD when using the 'strictest' of the four values for ϵ. This predictor represents the dimensionality of the network object if box-counting—the method for calculating FD—can only box together nodes which are both sufficiently close in links *and* within 0.25 of each other in (normalised) fitness. In this model, the coefficient is positive; this implies a higher value should be associated with a raised success rate by the ILS.

The SA model, in the middle column, is the strongest of the three, with over 70% of variance in the success rate explained. Here, three of the four FDs have statistical significance. Two are positive: the strictest value for ϵ, and the most lenient. The dimension with ϵ set at 0.5, however, has a negative coefficient— which is larger (in absolute value) than the other two.

If we now look at the TS summary, we can see from the R^2 that this is a very strong model. The fractal coefficients are positive without exception here. Two are noteworthy: the dimension at the most lenient value for ϵ, and the dimension with the value set at 0.5.

Table 3 summarises mixed models similar to those in Table 2, but this time with run-time as the dependent, instead of success rate. Again the predictors are described in Table 1. We can see from the R^2 that these are better models than when considering success rate, with over 90% of variance explained in all three cases.

The ILS model is of particular interest, with an R^2 of over 97%. The coefficients for the fractal predictors show the critical effect of the parameter ϵ. The coefficients are all negative, except for the most lenient value, which is also the only one with statistical significance. This implies a slower search in the case of a LON with high FD if considering mostly link-distance.

If we now look to the middle column in Table 3, we can see that the SA model is of very good fit. A similar trend to the ILS is seen: the fractal coefficients are negative, except for the one with largest value for ϵ, which is significant, alongside the dimension with it set at 0.5.

The final summary in the Table shows that the TS model is a extremely strong, with ∼98% of variance explained. Similarly to ILS and SA, the coefficients for the fractal predictors are negative in the case of three out of four. Again, the only exception is the dimension when using the largest value for ϵ. This predictor has a p-value which indicates significance; as does the dimension when using 0.5, which is larger in absolute value.

An important point in Tables 2 and 3 is that whenever fractal predictors had p-values indicating significance, their coefficients were orders of magnitude larger than those of the number of optima, or the number of funnels, even when these also had acceptable p-values.

5   Discussion

5.1   The Fractal Shape of Local Optima Networks

In Sect. 4.1, we saw that the instances under study have a varied dimension in their local optima networks.

The local optima networks which were extracted from instances with the lowest level of epistasis (and therefore ruggedness) had a FD interquartile range of ∼1.008 to ∼1.379. Clearly, most of these local optima networks are between one and two dimensional in the way they fill space. Having a dimension just above one implies a linear sort of structure, but with some extra pattern, such that it cannot be classified as a one-dimensional line. An interpretation of this for local optima networks is that the networks comprise a somewhat linear sequence of local optima, with a little bit of deviation or detail.

The networks from the $K = 4$ and 6 problems generally had higher dimensions. Some of these were complex shapes between one and two dimensions (for example, FD = 1.89), implying a line with a substantial amount of extra detail. This could mean that the networks are winding convoluted sequences, with spokes leading off the main path. Some networks from these categories had dimensions which were just above two. A fractional dimension just above an integer signifies a low-complexity object. For our local optima networks, this could possibly mean a low number of paths of local optima, with a bit of detail over and above 2d scaling.

The local optima networks which were extracted from the most rugged problems ($K = 8$) are virtually impossible to properly envision, as the majority of them had a FD greater than three. This being said, most were just above three, indicating the detail in the objects scales mostly in the manner of a three-dimensional object. It is possible that the three dimensions represent connecting or intersecting paths of local optima.

5.2   Connections with Search Difficulty

In Sect. 4.2 we saw that there are connections between dimension in a local optima network and search difficulty. When using $\epsilon = 1.0$, the calculated dimensions were linked to a raised success rate and a prolonged search time in metaheuristics. Both phenomena could be explained if we recall that many of the easier-instance LONs had dimension between one or two; in other words, the dimension implies a linear, sequence-like structure, with some degree of extra detail or pattern. The labyrinthine nature of the detail would explain the speed, and the one-dimensional base could mean that search algorithms on the underlying instance manage to get through the sequence of local optima to the global best.

Changing the value of the parameter ϵ changes the way the FDs are calculated and the results are markedly different. We saw in Table 3 that when values $\{0.25, 0.50, 0.75\}$ were used during the calculations, instead of 1.00, the corresponding network dimensions were associated with a quicker runtime in search algorithms. This is the opposite result to that obtained using $\epsilon = 1.00$. The difference lies in the boxing of nodes in the box-counting algorithm. With stricter criteria for nodes being boxed together, the network is partitioned in a different way. A high fractional dimension, therefore, might be associated with 'nearby' nodes (in edge-distance) not satisfying the fitness condition. Nearby nodes such as these might form paths with big fitness 'jumps' which might explain the quicker behaviour of search algorithms.

6   Conclusions and Future Work

We have conducted an empirical and introductory study on the potential for use of fractal analysis in local optima networks. A benchmark combinatorial problem, the NK Landscape Model, was used as a case study. Various extents of ruggedness were used.

The results indicate that when we consider our local optima networks to be general complex networks, a high FD relates to increased run-time for prominent metaheuristics, but also an enhanced rate of hitting the optimum. Accordingly, fractal analysis seems to capture a unique phenomenon in a local optima network: previously-proposed metrics have been linked to lowered performance only.

Another important result is the critical effect of the parameter specific to the case of a network composed of local optima, denoted here as ϵ. In particular, networks with high dimension when considering *fitness* as part of the scale used, were shown to be associated with increased efficiency in ILS and TS. We argue that this may be due to 'close' local optima being associated with large fitness jumps.

While the focus was on small problems here, there is no reason that fractal analysis could not be deployed on sampled local optima networks, given a robust sampling algorithm. With this new insight into the 'middle level' in a search space—the space of local optima—we can proceed further down this avenue of possibility. The more traditional (agnostic of the semantics of the network) fractal analysis captures a unique element of the local optima network: a phenomenon which is linked to lowered efficiency but raised effectiveness, and which we argue merits further investigation. Furthermore, the addition of the fitness-mandate condition for the special case of a local optima network gives valuable insight into the importance of optima connectivity and fitness distribution.

Acknowledgements. This work is supported by the Leverhulme Trust (award number RPG-2015-395) and by the UK's Engineering and Physical Sciences Research Council (grant number EP/J017515/1). We gratefully acknowledge that all network data used during this research were obtained from [2].

References

1. Weinberger, E.D., Stadler, P.F.: Why some fitness landscapes are fractal. J. Theor. Biol. **163**(2), 255–275 (1993)
2. Ochoa, G., Tomassini, M., Vérel, S., Darabos, C.: A study of NK landscapes' basins and local optima networks. In: Proceedings of the 10th Annual Conference on Genetic and Evolutionary Computation, pp. 555–562. ACM (2008)
3. Mandelbrot, B.: How long is the coast of Britain? Statistical self-similarity and fractional dimension. Science **156**(3775), 636–638 (1967). http://www.sciencemag.org/content/156/3775/636.abstract
4. Song, C., Gallos, L.K., Havlin, S., Makse, H.A.: How to calculate the fractal dimension of a complex network: the box covering algorithm. J. Stat. Mech: Theory Exp. **2007**(03), P03006 (2007)
5. Stadler, P.F.: Fitness landscapes. In: Lässig, M., Valleriani, A. (eds.) Biological Evolution and Statistical Physics. Lecture Notes in Physics, vol. 585, pp. 183–204. Springer, Heidelberg (2002). https://doi.org/10.1007/3-540-45692-9_10
6. Weinberger, E.: Correlated and uncorrelated fitness landscapes and how to tell the difference. Biol. Cybern. **63**(5), 325–336 (1990)
7. Locatelli, M.: On the multilevel structure of global optimization problems. Comput. Optim. Appl. **30**(1), 5–22 (2005)
8. Zelinka, I., Zmeskal, O., Saloun, P.: Fractal analysis of fitness landscapes. In: Richter, H., Engelbrecht, A. (eds.) Recent Advances in the Theory and Application of Fitness Landscapes. ECC, vol. 6, pp. 427–456. Springer, Heidelberg (2014). https://doi.org/10.1007/978-3-642-41888-4_15
9. Thomson, S.L., Daolio, F., Ochoa, G.: Comparing communities of optima with funnels in combinatorial fitness landscapes. In: Proceedings of the Genetic and Evolutionary Computation Conference, GECCO 2017, pp. 377–384. ACM, New York (2017). http://doi.acm.org/10.1145/3071178.3071211
10. Cahon, S., Melab, N., Talbi, E.G.: Paradiseo: a framework for the reusable design of parallel and distributed metaheuristics. J. Heuristics **10**(3), 357–380 (2004)

How Perturbation Strength Shapes the Global Structure of TSP Fitness Landscapes

Paul McMenemy$^{(\boxtimes)}$ ⓘ, Nadarajen Veerapen ⓘ, and Gabriela Ochoa ⓘ

Computing Science and Mathematics, University of Stirling, Stirling, UK
{paul.mcmenemy,nadarajen.veerapen,gabriela.ochoa}@stir.ac.uk

Abstract. Local optima networks are a valuable tool used to analyse and visualise the global structure of combinatorial search spaces; in particular, the existence and distribution of multiple funnels in the landscape. We extract and analyse the networks induced by Chained-LK, a powerful iterated local search for the TSP, on a large set of randomly generated (Uniform and Clustered) instances. Results indicate that increasing the perturbation strength employed by Chained-LK modifies the landscape's global structure, with the effect being markedly different for the two classes of instances. Our quantitative analysis shows that several funnel metrics have stronger correlations with Chained-LK success rate than the number of local optima, indicating that global structure clearly impacts search performance.

Keywords: Local optima network · Travelling salesman problem
Chained-LK · Perturbation strength · Combinatorial fitness landscape

1 Introduction

Characterising the global structure of combinatorial search spaces remains a challenge, partly due to the lack of tools to study their complexity. Local optima networks (LONs) [1,2] help to fill this gap by providing a modelling tool that compresses the fitness landscape into a network, where nodes are local optima and edges are possible search transitions between the local optima. Thus, LONs model the distribution and connectivity pattern of local optima. They help to characterise the underlying global landscape structure with a new set of metrics and visualisation tools [3].

Recent results using LONs challenge the existence of a 'big-valley' global structure on travelling salesman landscapes induced by iterated local search (ILS). Compelling evidence suggests instead that the big-valley decomposes into multiple valleys (or funnels) [4,5]. This multi-funnel structure helps to explain why ILS can quickly find high-quality solutions, but fails to consistently reach the global optimum.

The notion of a *funnel* was introduced within the protein folding community to describe "a region of configuration space that can be described in terms of

© Springer International Publishing AG, part of Springer Nature 2018
A. Liefooghe and M. López-Ibáñez (Eds.): EvoCOP 2018, LNCS 10782, pp. 34–49, 2018.
https://doi.org/10.1007/978-3-319-77449-7_3

a set of downhill pathways that converge on a single low-energy structure or a set of closely-related low-energy structures" [6]. It has been suggested that the energy landscape of proteins is characterised by a single deep funnel, a feature that underpins their ability to fold to their native state. In contrast, some shorter polymer chains (e.g., polypeptides) that misfold are expected to have other funnels that can act as traps. Similarly, recent studies on TSP [4,5] show that landscapes with more than one funnel, where the global optimum is located in a deep, narrow funnel, are significantly harder to solve for ILS.

Iterated local search is a simple yet powerful search strategy. It works by alternating an intensification stage (local search) with a diversification stage (perturbation). *Chained Lin-Kernighan* (Chained-LK) [7] is an effective ILS for TSP, combining the powerful Lin-Kernighan local search [8] with a type of 4-exchange perturbation (*double-bridge*, depicted in Fig. 1b). A key factor in any ILS implementation is the strength of the perturbation, which is related to the number of solution components that are modified simultaneously [9]. A recent study based on local optima networks from NK fitness landscapes shows that a properly selected perturbation strength can help overcome the effect of ILS becoming trapped in clusters of local optima (which are related to funnel structures) [10]. This has implications for the design of effective ILS approaches, where normally only small perturbations or complete restarts are applied, with the middle ground of intermediate perturbation strengths largely unexplored.

The main goal of this study is to model the LONs induced by Chained-LK on two classes of randomly generated TSP instances (Uniform and Clustered cities) for increasing perturbation strengths. More specifically, the contributions are:

1. A rigorous, empirical characterisation and comparison of the global structure of TSP instances with increasing perturbation strength.
2. Identification of the most effective perturbation strength for different TSP instance classes and sizes.
3. A correlation study identifying connections between empirical search performance and the global structure of TSP instances with increasing perturbation strength.

2   Definitions and Algorithms

The search space for a TSP instance with m cities is the set of permutations of legitimate Hamiltonian tours of these m cities; the number of tours is factorial in m. The fitness function f is given by the length of the tour, which is to be minimised for optimality.

The Lin-Kernighan (LK) heuristic is a powerful local search algorithm, one which is based on the idea of k-exchange moves: take the current tour and remove k different links; these are then reconnected in a new way to achieve a legal tour. A tour is considered to be 'k-opt' if no k-exchange exists which decreases its fitness value. LK applies 2, 3 and higher-order k-exchanges: Fig. 1a illustrates a 2-exchange move. The order of an exchange is not predetermined, rather k is increased until a stopping criterion is met. Thus many kinds of k-exchanges and all 3-exchanges are included.

Chained Lin-Kernighan (Chained-LK) is an iterated local search where the current solution obtained by LK is perturbed (or kicked) and LK is reapplied. If the new local optimum solution is an improvement, the old solution is discarded and the new one retained; otherwise, the search continues with the old tour and 'kicks' it again. The kick or escape operator in Chained-LK is a type of 4-exchange (depicted in Fig. 1b), named *double-bridge* by Martin et al. [11].

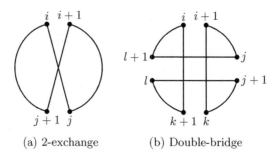

(a) 2-exchange   (b) Double-bridge

Fig. 1. Illustration of tours obtained after 2-exchange and double-bridge moves.

Fitness Landscape. A landscape [12] is a triplet (S, N, f) where S is a set of potential solutions, i.e., a search space; N is a neighbourhood structure, $N : S \longrightarrow 2^S$, a function that assigns to every $s \in S$ a set of neighbours $N(s)$; $f : S \longrightarrow \mathbb{R}$ is a fitness function.

Local Optima. A local optimum is a solution s^* such that $\forall s \in N(s^*)$, $f(s^*) \leqslant f(s)$. The set of local optima, which corresponds to the set of nodes in the network model, is denoted by L. Since the whole set of local optima cannot be determined in realistic search spaces, such as those considered here, a process of sampling is required to estimate L.

Escape Edges. These are directed edges, where there exists an *escape* edge from local optimum x to local optimum y, only if y can be obtained after a double-bridge move applied to x and is then followed by LK. The type of edge is dependent on the algorithm used to sample the landscape. While escape edges are natural edges for ILS algorithms, other edge types may be used such as crossover and mutation edges in the case of an evolutionary algorithm [13]. The set of escape edges obtained is denoted by E. Note that during the sampling process only edges that correspond to monotonically improving transitions are stored (see Algorithm 1).

Local Optima Network (LON). This is the graph $LON = (L, E)$ where all nodes are the local optima L, and edges E are the directed escape edges.

In some cases, neutrality may be observed at the LON level, i.e., there exists connected nodes that share the same fitness. This leads to an even coarser model of the landscape [14], where these LON plateaus are compressed into single nodes.

The set of LON plateaus, $CLp = \{clp_1, clp_2, \ldots, clp_n\}$ corresponds to the node set of a compressed local optima network. A node without neighbours that share the same fitness is also considered a LON plateau.

Compressed Local Optima Network ($CLON$). The weighted, oriented local optima network $CLON = (CLp, E)$ is the graph where the nodes $clp_i \in CLp$ are the LON plateaus. Weighted edges correspond to the aggregation of the multiple edges from nodes in a plateau to single edges in the compressed network.

Sink. A sink is a $CLON$ node without outgoing edges.

Funnel. The funnel of some sink is empirically defined as the set of nodes for which there exists a path of monotonically decreasing fitness between the nodes and the sink.

Data: I, TSP instance; z, the kick strength
Result: L, set of local optima; E, set of escape edges
$L \leftarrow \{\}; E \leftarrow \{\}$
for $i \leftarrow 1$ to 1000 do   /* loop across 1000 runs */
    $s_{start} \leftarrow$ LK(initialFeasibleSolution(I))
    $L \leftarrow L \cup \{s_{start}\}; j \leftarrow 0$
    while $j < 10000$ do   /* termination criterion of run */
        $s_{end} \leftarrow$ applyKick(s_{start}, z)   /* apply kick of strength z */
        $s_{end} \leftarrow$ LK(s_{end})
        $j \leftarrow j + 1$
        if $Objective(s_{end}) \leq Objective(s_{start})$ then
            $L \leftarrow L \cup \{s_{end}\}$ /* aggregate the nodes */
            $E \leftarrow E \cup \{(s_{start}, s_{end})\}$ /* aggregate the edges */
            $s_{start} \leftarrow s_{end}$
            $j \leftarrow 0$
        end
    end
end

Algorithm 1. LON sampling aggregating 1000 runs of Chained-LK.

3   Empirical Methodology

We empirically examine the global structure of two classes of synthetic instances by considering a sample of the search space. Clearly, exhaustive enumeration of the search space is not possible when instances of non-trivial sizes are considered. We therefore choose to sample high quality local optima obtained. Local optima networks were constructed using the Chained Lin-Kernighan heuristic as described above, and implemented with the Concorde TSP solver [15,16].

3.1   Instances

The instances considered here are generated using the DIMACS TSP instance generator[1]. Two classes of synthetic instances were generated: ones where the cities are uniformly distributed (prefixed by 'E') and ones where the cities are clustered (prefixed by 'C'). These two classes are further subdivided into three non-trivial instance sizes: 506, 755 and 1010 cities [4]. For each class-size combination, thirty different instances are generated; therefore a total of 180 instances is considered. A subset of these instances, and their associated networks, was created in the context of our previous work [4].

3.2   Sampling Method

The algorithm used to sample the search space is an instrumented version of the Chained-LK implementation found in Concorde [15,16]. Concorde is currently the state-of-the-art in exact TSP solvers and uses the Chained-LK heuristic to generate a good upper bound for its branch-and-bound process. In addition, we used Concorde to compute a global optimum (and its associated fitness value) for each instance.

The Chained-LK implementation is modified [4] to record all the local optima found during a run (set L), as well as the order in which the local optima were generated (set E), i.e., which local optimum was obtained after some other local optimum was perturbed and then improved through LK. The termination criterion of the algorithm is 10000 consecutive non-improving moves. The information for a total of 1000 runs is subsequently aggregated in order to generate the corresponding LON. The pseudocode is given in Algorithm 1.

For each instance, we separately sample the landscapes for ten perturbation strengths ($z \in \{1, 2, \ldots, 10\}$). The value of z is the number of consecutive double-bridge kicks that are applied in order to perturb a local optimum. The perturbations are therefore fairly disruptive, even at low strength. The total sampling process was relatively computationally expensive, requiring approximately 50000 hours of computation time on Intel Xeon X5650 2.66 GHz processors.

---

[1] dimacs.rutgers.edu/Challenges/TSP/download.html.

3.3   Performance and Network Metrics

There are multiple performance and network metrics that can be computed and used to understand search difficulty and landscape structure; however, not all of them are relevant. We explored a number of metrics and selected a subset that allows us to meaningfully describe the relationship between search difficulty and landscape structure. Table 1 summarises each of these metrics, and which are described in some additional detail below.

Table 1. Definitions of performance and network metrics

Performance metrics	$success$	Success rate of finding the *A priori* global optimum
	$iters$	Mean no. of Chained-LK iterations to find global optimum
Network metrics	n	Number of local optima within the LON
	n_{sinks}	Number of sinks within the LON
	d	Relative in-strength of the globally optimal sinks
	fun	Proportion of nodes comprising the globally optimal funnels
	u_{fit}	Proportion of local optima with unique fitness values

Algorithmic performance is measured through its *success* rate; i.e., the proportion of runs that reached a global optimum, and through the mean number of iterations (*iters*) required to reach a global optimum.

The structure of a LON is assessed using five characteristics. The number of local optima (n), or nodes, within the LON describes its size. A sink describes a solution in which the search becomes trapped at the end of a funnel; the corresponding metric we record is n_{sinks}, the number of sinks.

In addition, the weighted incoming degree, or in-strength, of each sink is computed and the relative in-strength of the globally optimal sinks (d) is reported. This corresponds to the ratio of the sum of the in-strengths of the globally optimal sinks to the sum of the in-strengths for all sinks.

The nodes in a funnel, in the uncompressed LON, are identified by selecting one of the nodes that constitutes a sink and performing a breadth-first search. We thus define *fun* as the proportion of nodes of the LON within funnels that lead to a global optimum.

Finally, u_{fit} is the proportion of local optima with unique fitness values, i.e., the number of unique fitness values divided by the total number of unique local optima.

4   Results and Analysis

In this section, we examine the structure of the sampled LONs and the inter-play with search difficulty. This is initially achieved through visualisation, to intuitively reveal structural differences; and then through a more traditional analysis of the performance and network metrics.

4.1   Visualisation

One advantage of modelling fitness landscape as LONs is the possibility of net-work visualisation, a valuable first approach in exploring and understanding their structure. Figure 2 illustrates LONs for two representative instances with 755 cites. Specifically, we considered one Uniform (E755.81) and one Clustered (C755.81) instance, and visualise the networks for three different perturbation strengths, $z \in \{1, 5, 10\}$. Each node corresponds to an LK optimum, and edges represent search transitions with the corresponding perturbation strength. Note that other instances were examined, and produced similar results.

In order to produce images of manageable size, each plot represents a sub-network of the sampled network. Nodes were kept whose fitness was within 5% of the evaluation of the global optimum. This visualises the connectivity pattern of solutions nearby the global optimum, which is arguably the most interesting part of the search space, and any competent heuristic method should attain this portion of the landscape.

Network plots were produced using the R statistical language together with the igraph package [17]. Graph layouts employ *force-directed* methods, and resul-tant visualisations are decorated to reflect features relevant to search dynamic. It is important to remember that a LON can be seen as a representation of the stochastic process of a search algorithm for a particular problem instance. In Fig. 2, red nodes correspond to local optima belonging to the funnel containing the global optimum, whereas blue nodes indicate optima belonging to suboptimal funnels. The bottom of the funnels (sinks) are highlighted with a black outline and a node size proportional to their incoming strength (weighted degree). Nodes which are not sinks are visualised with a grey outline and fixed size.

A visual inspection of the LONs in Fig. 2 reveals clear structural differences between the global shape of the studied landscapes, as well as the perturba-tion strengths. For the Uniform instances (left plots), the global structure and overall success rate of Chained-LK seems very robust to increasing perturba-tion strength. Interestingly, the success rate achieved with the standard single double-bridge kick ($z = 1$) is significantly lower than that achieved when $z = 5$. This can be explained by the existence of a suboptimal funnel seen in blue (Fig. 2(a)) which is disjoint from the global sink in red, and whose sink in blue has incoming strength of similar magnitude (visualised as the size of the node). Therefore this suboptimal funnel acts as a trap for the search process. Once the algorithm reaches the blue sink, it cannot escape using a single kick and thus fails to locate the global optimum. The intermediate kick strength ($z = 5$) shown in Fig. 2(c) provides the best success rate for this instance, supported by the single

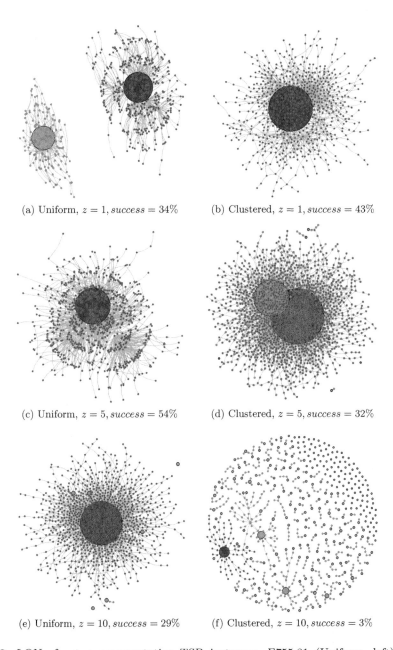

(a) Uniform, $z = 1, success = 34\%$ (b) Clustered, $z = 1, success = 43\%$

(c) Uniform, $z = 5, success = 54\%$ (d) Clustered, $z = 5, success = 32\%$

(e) Uniform, $z = 10, success = 29\%$ (f) Clustered, $z = 10, success = 3\%$

Fig. 2. LONs for two representative TSP instances, E755.81 (Uniform, left) and C755.81 (Clustered, right), and three perturbation strengths $z \in \{1, 5, 10\}$. *success* percentages of Chained-LK are indicated. Nodes are LK local optima and edges are transitions using the respective perturbation strength. Red nodes belong to the global optimal funnel, while blue nodes to suboptimal funnels. The size of sink nodes is proportional to their incoming strength. (Color figure online)

funnel observed. Thus the $z = 5$ perturbation seems to fuse the previous two funnels into one. However, too strong a perturbation deteriorates performance (Fig. 2(f)), which can be explained by the appearance of several suboptimal sinks (blue nodes), albeit each of small incoming strength.

4.2 Performance and Network Metrics Results

The metrics previously defined in Table 1 are now examined in detail. Figure 3 provides boxplots of the metric values for the six TSP class-size instances. The metrics are shown grouped by class and are vertically descending by size (506, 755, 1010). Results for each metric presented in Fig. 3 are discussed hereafter:

success. A significant difference is observed between the distributions of *success* for each class-size combination. The *success* rates of Uniform instances for each of the three sizes can be approximately characterised as concave, that is they follow an increasing then decreasing behaviour as z increases. However, the Clustered results all decay from maximum values of *success* when z is low, diminishing towards zero as $z \to 10$. The variability of *success* results also narrow as instance size increases for both instance classes.

iters. For each class-size combination, an approximately monotonic increase in the iterations of Chained-LK is observed as the strength of z increases. However, the Clustered 1010 results peak when $z = 7$; the *iters* begin to decrease when $z \geqslant 8$. This is a consequence of the very low success rate for these instances, which is close to zero for these values of z. When Chained-LK does obtain the global optimum for these instances, it does so quickly due to the small size of the globally optimal funnel (c.f. Fig. 2(f)). The Uniform classes all exhibit lower *iters* values almost without exception when compared with the Clustered classes; only when z is low is there any overlap in the Uniform and Clustered interquartile boxes between comparable results.

d. The relative in-strengths of the funnels leading to the global optima exhibit similar behaviours to that of the corresponding success rates for both Uniform and Clustered classes. This is particularly apparent when comparing the boxplots of *success* and d for instance size 1010. Note that the magnitude of d decreases as instance size increases.

n. All plots in Fig. 3 exhibit values of local optima that decrease as kick strength increases. As detailed in the sampling method (Sect. 3.2), a perturbation away from the current local optimum employs z double-bridge moves; as z increases, this process becomes increasingly destructive of the TSP tour relevant to the current local optimum. This results in the perturbation stage relocating the search to a location on the fitness landscape that is increasingly distant from the current solution. This results in Chained-LK becoming less likely to discover a non-deteriorating solution after each iteration. Thus, the number of new local optima generated would reduce as z increased. Note that the Clustered instances again return greater numbers of local optima when compared to the Uniform results.

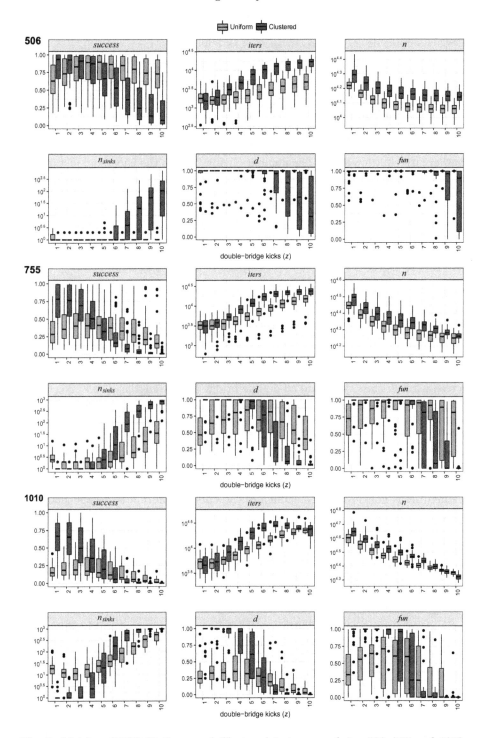

Fig. 3. Metrics of TSP Uniform and Clustered instances, of size 506, 755 and 1010, with increasing double-bridge kicks (z). Note the $\log_{10}$ vertical scale on some of the plots.

fun. As with d, the proportional size of a funnel containing the global optimum follows a similar decay pattern to *success* as z increases.

n_{sinks}. We first consider the number of sinks for the Uniform class as z increases. When instance size is 506, only a small number of sinks are observed; however, as instance size increases to 755 and 1010, the value of n_{sinks} also increases. In actuality, the number of sinks decreases from $z = 1$ to $z = 3$, and then begins to increase as $z \to 10$. These results are in accordance with the visualisations in Fig. 2, which also demonstrate that small perturbation strengths could meld together large but few sinks. The number of sinks within the Clustered instances all follow a logistic growth pattern as perturbation strength z increases. Both instance classes approximate reciprocal n_{sinks} behaviours to that of the performance metric *success*; therefore it is expected that there would exist a strong, negative correlation between these metrics.

4.3   Impact of Perturbation Strength on Success Rate

Considered together, the three upper left plots for each size shown in Fig. 3 provide indicators as to how the *success* rates alter as TSP instance size increases. The median values from the distributions of *success* are reproduced in Fig. 4; graphed by type (Uniform and Clustered), and then as connected lines by instance size (506, 755, 1010).

It is apparent from Fig. 4 that, as instance size increases, the rate of *success* decreases for all class types and all double-bridge kick strengths. The E506 (Uniform) median *success* rates all exhibit relatively large values, ranging from a high of 89.5% for kick strength $z = 5$ to a lowest rate of 62.9% when $z = 1$. This could be a consequence of using too low a perturbation z, one that is not strong enough to kick sufficiently 'far' away from the current optimum and so failing to escape to a new, non-deteriorating local optimum, even if the current optimum is in a relatively shallow funnel. Thus, for low values of z, the perturbation stage of Chained-LK is not destructive enough of the current solution to escape shallow traps, and continue on to find the global optimum.

A marked decrease in the median *success* values are observed for Uniform TSP instance sizes increasing from 755 to 1010. This increase is primarily a consequence of the increased magnitude of the solution space. The maximum median *success* for each of E506, E755, E1010 occur when double-bridge kick strengths are at $z = 5$ (median = 0.895), $z = 5$ (median = 0.410) and $z = 3$ (median = 0.190) respectively.

Similar analysis of the Clustered TSP instances shows notably different behaviour in the values of median *success* values for $z \in \{1, 2, \ldots, 10\}$. Where the Uniform TSP instances' median *success* values exhibit an approximately concave shape, the Clustered median *success* values conform to a complemented sigmoid pattern. This locates each Clustered instance's maximum, median value when $z = 1$; the median *success* values then decay towards zero as $z \to 10$.

As TSP instance size increases, both Uniform and Clustered classes experience a significant reduction in the median *success* rates, as well as a phase shift

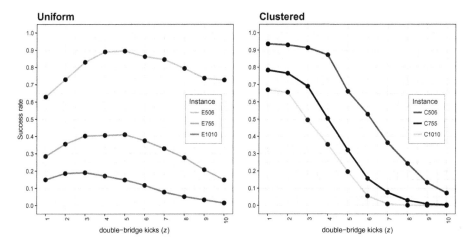

Fig. 4. Success rates by TSP instance class for increasing instance sizes. Points shown are medians of *success* for each class-kick combination.

of their patterns towards the vertical axis. Extrapolating from these behaviours we would reasonably expect that, as TSP instance size increases beyond 1010, *success* rates would continue to degrade; maximum, median *success* would thus occur for lower values of z. Conversely, for smaller TSP instance sizes, *success* rates would logically approach 100%, with close to 100% *success* rates expected for reasonable values of z as TSP size decreased.

4.4   Correlation Analysis

Analysis of correlations between the performance metric *success* and the other metrics *iters*, n, n_{sinks}, d, *fun* and u_{fit} was undertaken. Figure 5 shows the correlation matrix for these metrics (for C755, or Clustered TSP instances of size 755, only), with the upper triangle of the matrix providing Spearman correlation coefficients for each metric pair combination. Asterisks next to each correlation coefficient denote the p-value significance, where *** indicates a p-value < 0.001, ** a p-value < 0.01, and a single * denotes p-value < 0.05. The lower triangle of Fig. 5 shows pairwise scatter plots of each metric-metric combination. Pairwise scatter points shown are aggregated together from each instance and kick. Correlations matrices for all 6 type-size combinations were produced but not shown; however, the correlations for C755 are indicative of all instances, and so is the only matrix shown here for brevity.

We are primarily interested in correlations between *success* and the other metrics, and so need only consider coefficients shown in the first row of Fig. 5 and scatter plots within its first column. In the first row, we observe high positive correlations between *success* and the network metrics d and *fun*: the relative in-strength and funnel sizes of the global optima respectively. Strong negative correlations with *success* are seen with *iters* (iterations to global optimum) and

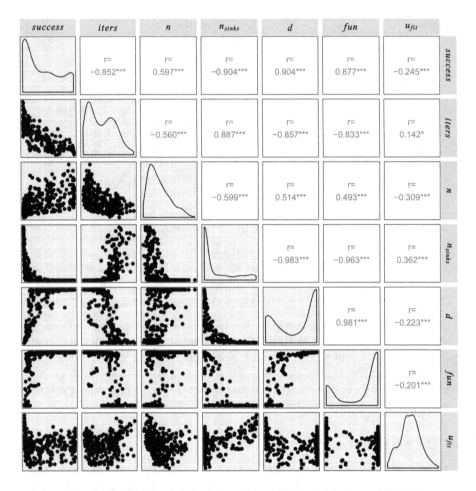

Fig. 5. Correlation matrix for combined Clustered instances C755.

n_{sinks} (number of sinks). Only *success* correlations with number of local optima (n) and the proportion of nodes with unique fitness (u_{fit}) exhibit coefficient values that could be described as being of moderate significance.

Note that all correlation coefficients provided here were calculated using Spearman's approach rather than Pearson's; thus coefficients reflect correlations based on monotonically increasing or decreasing relationships rather than linear interactions only. However, all correlations were recalculated using Pearson's methodology, and no significant differences can be reported except for an across-the-board small reduction in the magnitudes of the coefficients.

4.5   Correlation Variance Between Instance Classes

Correlations of success rates in finding a global optimum (*success*) with the other metrics are grouped by class (Uniform, Clustered) and by TSP size and are shown as parallel plots in Fig. 6. This allows easy comparison of *success* correlates between the TSP classes. For both Uniform and Clustered classes, similar behaviours transpire for each of the metrics correlated to *success*. Increasing from 506 to 755 then to 1010 TSP sizes, we observe positive, strengthening *success* correlations with d, *fun*; strengthening, negative correlations are observed with *iters* and n_{sinks}. Both the number of local optima (n) and neutrality at the local optima level (u_{fit}) undergo changes in the directions of their correlation as TSP size increases, both transiting coefficient values of zero and thus emphasising their weak correlation with *success* rate.

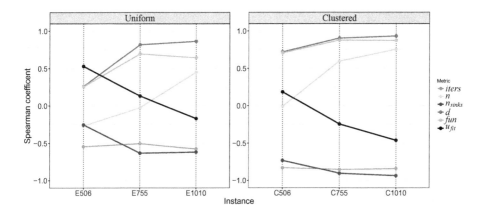

Fig. 6. Spearman correlation coefficient values of *success* versus other metrics by class.

Extrapolating these correlation behaviours beyond the TSP sizes we have analysed here would be inappropriate. We observed an unexpected decrease in the number of iterations for the C1010 instance (c.f. Fig. 3) as $z \to 10$, something that could not have been readily anticipated from the C506 and C755 results. Therefore, making assumptions about trends from these results as TSP size increases would not be advisable; studies of larger TSP sizes would be required to further understand the relationships between algorithm performance and LON structure.

5   Conclusions

We have extracted and analysed the local optima networks induced by Chained-LK with increasing perturbation strengths on two classes of randomly generated TSP instances, with Uniform and Clustered cities. Particularly, we characterised, visualised and measured the multi-funnel global structure of the underlying fitness landscapes. We have also measured the performance of Chained-LK on the studied instances with increasing perturbation strengths. Our results reveal that the landscape global structure (i.e., the funnel structure) changes with increasing perturbation strength, and these changes are markedly different in the two instance classes. For the Clustered instances, a smoother landscape (a single or a low number of funnels) is achieved with a minimal perturbation, whereas a strong perturbation is ineffective and breaks the landscape into a multitude of suboptimal funnels. In contrast, for the Uniform instances, the lowest perturbation induces a multi-funnel landscape, while a moderate perturbation can smooth the landscape and improve performance. This has implications for algorithm design: the amount of perturbation for an effective search is related to the instance class. However, we found that the standard perturbation strength of Chained-LK (a single double-bridge move) is robust across the instance classes and sizes studied.

Provided the most effective perturbation strength is selected for each instance class, the Clustered instances are generally easier to solve (have a higher success rate) across all instances sizes. However, the Clustered instances reveal a larger number of local optima according to our sampling, as compared to the Uniform instances across all sizes. If the number of local optima was an accurate predictor of search difficulty, we would expect that the Clustered instances would be harder to solve than the Uniform instances. The opposite is the case, which confirms that the distribution of optima is more important than their number for explaining search difficulty. Indeed, we found that the funnel metrics (the number of funnels, the size and strength of the global optima funnel) have a stronger correlation with search difficulty than the number of local optima. Global structure definitely impacts search performance.

Future work will consider other classes of TSP instances including real-world datasets. Additional problem domains will also be considered, as well as the impact on global structure of alternative perturbation and recombination operators.

Acknowledgements. This work was supported by the Leverhulme Trust [award number RPG-2015-395] and by the UK's Engineering and Physical Sciences Research Council [grant number EP/J017515/1]. Results were obtained using the EPSRC-funded ARCHIE-WeSt High Performance Computer (www.archie-west.ac.uk, EPSRC grant EP/K000586/1).
Data Access. All data generated for this research are openly available from the Stirling Online Repository for Research Data (http://hdl.handle.net/11667/104).

References

1. Ochoa, G., Tomassini, M., Verel, S., Darabos, C.: A study of NK landscapes' basins and local optima networks. In: Proceedings of Genetic and Evolutionary Computation Conference (GECCO), pp. 555–562. ACM (2008)
2. Verel, S., Ochoa, G., Tomassini, M.: Local optima networks of NK landscapes with neutrality. IEEE Trans. Evol. Comput. **15**(6), 783–797 (2011)
3. Newman, M.E.J.: Networks: An Introduction. Oxford University Press, Oxford (2010)
4. Ochoa, G., Veerapen, N.: Mapping the global structure of tsp fitness landscapes. J. Heuristics 1–30 (2017). https://doi.org/10.1007/s10732-017-9334-0. ISSN 15729397

5. Ochoa, G., Veerapen, N.: Deconstructing the big valley search space hypothesis. In: Chicano, F., Hu, B., García-Sánchez, P. (eds.) EvoCOP 2016. LNCS, vol. 9595, pp. 58–73. Springer, Cham (2016). https://doi.org/10.1007/978-3-319-30698-8_5
6. Doye, J.P.K., Miller, M.A., Wales, D.J.: The double-funnel energy landscape of the 38-atom Lennard-Jones cluster. J. Chem. Phys. **110**(14), 6896–6906 (1999)
7. Applegate, D., Cook, W., Rohe, A.: Chained Lin-Kernighan for large traveling salesman problems. INFORMS J. Comput. **15**, 82–92 (2003)
8. Lin, S., Kernighan, B.W.: An effective heuristic algorithm for the traveling-salesman problem. Oper. Res. **21**, 498–516 (1973)
9. Lourenço, H.R., Martin, O.C., Stützle, T.: Iterated local search. In: Handbook of Metaheuristics, pp. 320–353. Kluwer Academic Publishers, Boston (2003)
10. Herrmann, S., Herrmann, M., Ochoa, G., Rothlauf, F.: Shaping communities of local optima by perturbation strength. In: Genetic and Evolutionary Computation Conference, GECCO, pp. 266–273 (2017)
11. Martin, O., Otto, S.W., Felten, E.W.: Large-step Markov chains for the TSP incorporating local search heuristics. Oper. Res. Lett. **11**, 219–224 (1992)
12. Stadler, P.F.: Fitness landscapes. Appl. Math. Comput. **117**, 187–207 (2002)
13. Veerapen, N., Ochoa, G., Tinós, R., Whitley, D.: Tunnelling crossover networks for the asymmetric TSP. In: Handl, J., Hart, E., Lewis, P.R., López-Ibáñez, M., Ochoa, G., Paechter, B. (eds.) PPSN XIV. LNCS, vol. 9921, pp. 994–1003. Springer, Cham (2016). https://doi.org/10.1007/978-3-319-45823-6_93
14. Ochoa, G., Veerapen, N., Daolio, F., Tomassini, M.: Understanding phase transitions with local optima networks: number partitioning as a case study. In: Hu, B., López-Ibáñez, M. (eds.) EvoCOP 2017. LNCS, vol. 10197, pp. 233–248. Springer, Cham (2017). https://doi.org/10.1007/978-3-319-55453-2_16
15. Applegate, D.L., Bixby, R.E., Chvátal, V., Cook, W.J.: The Traveling Salesman Problem: A Computational Study. Princeton University Press, Princeton (2007)
16. Applegate, D., Bixby, R., Chvátal, V., Cook, W.: Concorde TSP solver (2003). http://www.math.uwaterloo.ca/tsp/concorde.html
17. Csardi, G., Nepusz, T.: The igraph software package for complex network research. Int. J. Complex Syst. **1695**, 1–9 (2006)

Worst Improvement Based Iterated Local Search

Sara Tari$^{(\boxtimes)}$, Matthieu Basseur, and Adrien Goëffon

Laboratoire d'Etude et de Recherche en Informatique d'Angers,
UFR Sciences, 2 Boulevard Lavoisier, 49045 Angers Cedex 01, France
{sara.tari,matthieu.basseur,adrien.goeffon}@univ-angers.fr

Abstract. To solve combinatorial optimization problems, many metaheuristics use first or best improvement hill-climbing as intensification
mechanism in order to find local optima. In particular, first improvement
offers a good tradeoff between computation cost and quality of reached
local optima. In this paper, we investigate a worst improvement-based
moving strategy, never considered in the literature. Such a strategy is
able to reach good local optima despite requiring a significant additional
computation cost. Here, we investigate if such a pivoting rule can be
efficient when considered within metaheuristics, and especially within
iterated local search (ILS). In our experiments, we compare an ILS using
a first improvement pivoting rule to an ILS using an approximated version of worst improvement pivoting rule. Both methods are launched
with the same number of evaluations on bit-string based fitness landscapes. Results are analyzed using some landscapes' features in order to
determine if the worst improvement principle should be considered as a
moving strategy in some cases.

1 Introduction

Over the last two decades, the number of new metaheuristics in the literature
has been unreasonably growing. Although the creativity of researchers can be
outstanding, it is not conceivable to create a new paradigm for each considered optimization problem. Moreover, it is hard to choose the most adequate
metaheuristic to tackle a given optimization problem among the great variety of
existing methods. Indeed, many metaheuristics are proposed for a single problem
without a further analysis of their behavior. As stated by Sörensen in [1], there
is a need to understand metaheuristics components in order to know when and
how to use them. Nevertheless, there exists some studies which try to obtain
insights about metaheuristics behavior, some of them using the concept of fitness landscapes. Some previous works are particularly dedicated to hill-climbing
algorithms (climbers) and will be discussed in Sect. 3.

At first, focusing on climbers may seem obsolete since there exists metaheuristics more efficient to handle optimization problems. Yet, climbers are especially
basic and often used as intensification mechanisms of more sophisticated metaheuristics. We believe there is a need to deconstruct metaheuristics in order to

© Springer International Publishing AG, part of Springer Nature 2018
A. Liefooghe and M. López-Ibáñez (Eds.): EvoCOP 2018, LNCS 10782, pp. 50–66, 2018.
https://doi.org/10.1007/978-3-319-77449-7_4

obtain insights about their behavior and one way of doing this is to study some basic components of metaheuristics, including intensification techniques.

In this work, we focus on a climbing technique based on the worst improvement pivoting rule, which is a counter-intuitive moving strategy since it selects the least improving neighbor at each step of the search. Such a strategy showed interesting results in a previous work [2], since it generally allows the attainment of better local optima than when using first improvement. Yet, such a method needs a higher computational budget, even if worst improvement approximation variants offer interesting tradeoffs between quality and computation cost. Here, we propose to investigate further the behavior of worst improvement and approximated variants. First, we provide an extended analysis of such a climber, thanks to new experiments also combined with landscape analysis. Then, we compete iterated local searches using first improvement and worst improvement variants climbers, in order to determine their relative efficiency in an iterated context. Indeed, even if first improvement is outperformed by worst improvement in terms of local optima quality, its low computation cost allow to perform more climbing processes in a fixed maximal number of evaluations.

This study is conducted on two different types of bit-string fitness landscapes: NK landscapes and UBQP landscapes (derived from Unconstrained Binary Quadratic Programming problem instances).

In the next section we introduce definitions of concepts and problems used in this study. Section 3 is devoted to motivations of our work and previous results. In Sect. 4 we report and analyze results by means of some combinatorial landscape properties. Finally, in the last section we discuss our work.

2 Definitions

The concept of fitness landscape was originally introduced by Wright [3] in the field of theoretical biology, to represent an abstract space which links individual genotypes with their reproductive success. In particular, fitness landscapes are useful to simulate mutational paths and to observe the effect of successive mutations. Nowadays, such a concept is used in various fields to study the behavior of complex systems. In evolutionary computation, fitness landscapes can help to understand and thus to predict the behavior of neighborhood-based solving algorithms regardless of the problem considered. Let us introduce some concepts and definitions, then we will present fitness landscape instances used in this paper.

2.1 Fitness Landscapes and Related Concepts

A fitness landscape is formally defined by a triplet $(\mathcal{X}, \mathcal{N}, f)$ where $\mathcal{X}$ denotes the set of feasible solutions also called search space, $\mathcal{N}$ denotes the neighborhood function which assigns a set of neighboring solutions to each solution of the search space and f denotes the fitness function which assigns a value to each solution.

Fitness landscapes are determined by many characteristics relative to their topology, ruggedness, neutrality, etc. In the current work we mainly focus on two classical features: size and ruggedness.

In the following the size of a combinatorial fitness landscape will denote its number of solutions. The ruggedness is directly linked to the number and the repartition of the local optima. A landscape with a few local optima and large basins of attraction can be views as *smooth* whereas a landscape with many local optima and small basins is *rugged*. The basin of attraction of a local optimum [4] refers to the set of solutions from which a basic hill-climbing algorithm has a substantial probability to reach the considered local optimum. The presence of several local optima is directly related to the epistasis phenomenon. Epistasis occurs when the presence or absence of a given gene influences the fitness variation induced by a mutation. Sign epistasis occurs when a specific mutation improves a solution and deteriorates another one, as depicted in Fig. 1. If the two solutions where the mutation is applied are neighbors we refer to 1-sign-epistasis, if the distance between these solutions is of k we refer to k-sign-epistasis [5]. In an iterative improvement context, we mostly refer to sign epistasis since such methods mostly focus on the improving and deteriorating aspect of moves.

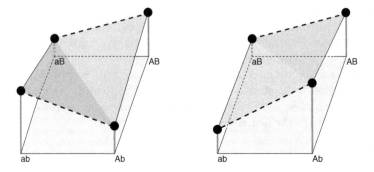

Fig. 1. Left-hand side: the presence of a mutation $a \rightarrow A$ affects the effect of the mutation $b \rightarrow B$ which becomes beneficial, there is sign-epistasis. Right-hand side: no sign epistasis.

For most combinatorial fitness landscapes it is not conceivable to enumerate all solutions, thus many landscape properties such as ruggedness cannot be exhaustively calculated. Many indicators can be used in order to estimate landscape properties [6], especially ruggedness. Here, we focus on the k-ruggedness indicator formally introduced in [5] and based upon k-sign epistasis. k-ruggedness refers to the k-epistasis rate within a sampling of pairs of k-distant solutions. 1-ruggedness which refers to the 1-epistasis rate on several pairs of mutation reflects local ruggedness whereas k-ruggedness (with $k > 1$) refers to a more global ruggedness of the landscape.

2.2   Bit-String Landscapes Instances

The NK landscape model [7] is widely used to generate artificial combinatorial landscapes with tunable size and ruggedness. NK landscapes are regularly used to understand the behavior of evolutionary processes in function of ruggedness levels.

Properties of NK landscapes are determined by means of two parameters N and K. N refers to the length of bit-string solutions and $K \in \{0, \ldots, N-1\}$ specifies the variable interdependency hence the ruggedness of the landscape. Setting K to 0 leads to a completely smooth landscape consequently containing a single local (then global) optimum. Increasing K induces an increase in the ruggedness of generated landscapes.

The topology of an NK landscape is a N-dimensional hypercube whose vertices are solutions and edges are neighborhood transitions (1-flip moves). The fitness function f_{NK} to be maximized is defined as follows:

$$f(x)_{NK} = \frac{1}{N} \sum_{i=1}^{N} C_i(x_i, \Pi_i(x))$$

x_i is the i-th bit of the solution x. Π_i is a subfunction which defines the dependencies of bit i, with $\Pi_i(x)$ such that $\pi_j(i) \in \{1, \ldots, N\} \setminus \{i\}$ and $| \cup_{j=1}^{N} \pi_j(i)| = K$. Subfunction $C_i : \{0,1\}^{K+1} \rightarrow [0,1)$ defines the contribution value of x_i w.r.t. its set of dependencies $\Pi_i(x)$. NK landscapes instances are determined by the $(K+1)$-uples $(x_i, x_{\pi_1(i)}, \ldots, x_{\pi_K(i)})$ and a matrix C of fitness contribution which describes the $2^N \times (K+1)$ possible contribution values.

The Unconstrained Binary Quadratic Programming problem (UBQP) is an NP-hard problem [8] regularly used to reformulate a large scope of real-life problems. An instance of UBQP is composed of a matrix Q filled with $N \times N$ positive or negative integers. A solution of UBQP is a binary vector x of size N where $x_i \in \{0,1\}$ corresponds to the i-th element of x. The objective function f_{UBQP} to be maximized is described as follows:

$$f_{UBQP}(x) = \sum_{i=1}^{n} \sum_{j=1}^{N} q_{ij} x_i x_j$$

The considered UBQP neighborhood is defined by the N solutions being at a Hamming distance of 1 (i.e. due to one-flip neighborhood operator). Thus N-dimensional NK and UBQP landscape instances only differ by their fitness function.

In experimental sections, a sample of 28 NK and 24 UBQP landscapes is used. NK landscapes are randomly generated using different size and variable interdependency parameter values: $N \in \{128, 256, 512, 1024\}, K \in \{1, 2, 4, 6, 8, 10, 12\}$. For UBQP landscapes, we use a set of 24 UBQP random instances[1], using different size and density parameter values: $N \in \{2048, 4096\}, d \in \{0.10, 0.25, 0.50, 0.75, 1\}$. The density $d \in [0,1]$ represents the expected proportion of non-zero values in the matrix Q.

---

[1] UBQP instances have been obtained with the instance generator provided at http://www.personalas.ktu.lt/~ginpalu/ubqop_its.html.

3 Worst Improvement Hill-Climbing

3.1 Pivoting Rules

A hill-climbing algorithm (climber) consists of selecting an improving neighbor of the current solution at each step of the search. When the current solution has no improving neighbor, a local optimum is reached and then implies the end of the climbing process.

The main interrogation when designing a climber concerns the neighbor selection rule (*pivoting rule*), which directly affects the capacity of the search algorithm to reach good local optima. This is particularly true since the global optimum (or at least a high local optimum) can actually be reached using a single climber from most solutions, assuming the use of a relevant pivoting rule [9].

Several works about climbers [2,10,11] investigate the effects of using classical first and best improvement pivoting rules on the capacity of climbers to reach good solutions. These studies highlight that first improvement is often the most efficient rule to reach good local optima on sufficiently large and rugged landscapes. Let us recall that first improvement (F) consists of selecting the first improving neighbor encountered at each step of the search whereas best improvement (B) selects the improving neighbor with the highest fitness value at each step of the search. Since performing smaller steps with first improvement often leads toward better local optima, a study considered the worst improvement (W) pivoting rule [2], which consists of selecting the improving neighbor with the lowest fitness value at each step of the search. This work showed that W is more efficient than F and B on non-smooth NK landscapes (some results are extracted on the left side of Table 1).

Table 1. Extract of previous results obtained with various hill-climbing on NK landscapes from [2]. For each couple (landscape,climber) we report the average fitness value of local optima as well as the average number of evaluations underneath.

Landscape	F	B	W	W_2	W_4	W_8	W_{16}
$NK_{256,4}$	.7128	.7211	**.7274**	.7233	.7254	.7257	.7262
	2k	19k	136k	5k	10k	21k	41k
$NK_{256,8}$	.7179	.7147	**.7267**	.7218	.7243	.7259	.7267
	2k	13k	284k	13k	31k	73k	66k
$NK_{256,12}$	.7053	.7015	**.7129**	.7089	.7105	.7124	.7122
	2k	10k	346k	5k	13k	33k	74k
$NK_{1024,4}$	.7238	.7232	**.7298**	.7253	.7270	.7286	.7291
	12k	302k	2336k	25k	54k	122k	223k
$NK_{1024,8}$	.7215	.7176	**.7330**	.7251	.7285	.7306	.7316
	13k	214k	5837k	31k	74k	172k	7330k
$NK_{1024,12}$	.7107	.7064	**.7210**	.7150	.7178	.7197	.7206
	14k	166k	8544k	34k	84k	206k	487k

Worst improvement requires a very high computational budget in terms of number of evaluations since it evaluates the whole neighborhood at each step of the search and generally performs more steps than first and best improvement. To overcome this issue, intermediate rules W_k have been proposed in [2]. W_k approximates W by selecting the solution with the lowest fitness among k improving neighbors at each step of the search. This process avoids the generation of the whole neighborhood at each step of the search and can drastically reduce the required number of evaluation to attain a local optimum. Some results of the aforementioned study are extracted in the right side of Table 1. W_k leads toward better solutions than first and best improvement even with $k = 2$ while significantly reducing the number of evaluations during the hill-climbing process. Let us notice that smooth NK landscapes are not considered since worst improvement is not efficient on such landscapes in comparison to a best-improvement hill-climbing.

3.2   Additional Experiments

In the current study we first perform additional experiments on UBQP landscapes (see Sect. 2.2), following the same protocol than in [2]. For each couple (landscape, method) we perform 100 runs starting from the same set of 100 randomly generated solutions. Results show that on UBQP landscapes (Table 2, left side), worst improvement always leads toward better local optima averages than first and best improvement but requires a huge number of fitness evaluations. Approximated versions of worst improvement are also considered (Table 2 - right side) and W_k appears to be particularly efficient while drastically reducing the number of evaluations. These results are similar to those obtained on NK landscapes and confirm the potential interest of worst improvement and its approximated variants.

The worst improvement rule was initially described in order to obtain better insights on what makes a climber efficient. Results showed that, despite the high computation cost, performing small improvements drives toward high local optima. Then, worst improvement approximations were proposed to obtain efficient pivoting rules which are not time consuming and thus potentially usable when tackling an optimization problem.

First improvement is the fastest and simplest way to reach local optima, which is an advantage when considering a fixed number of evaluations since it is able to reach a maximal number of local optima compared to any other strict hill-climbing pivoting rule. Considering more advanced pivoting rules which are more efficient but also more time consuming have to be investigated in a iterated context. Indeed, it is relevant to wonder if reaching less local optima than with first improvement is counterbalanced by their higher quality. In particular, we investigate if iterating W_k can be a better alternative than iterating first improvement in some cases. Let us recall that W_k leads toward better solutions than a climber using a first improvement pivoting rule, but it still requires at least twice more evaluations.

Table 2. Average fitness values (rounded to the nearest integers) of local optima obtained from 100 runs for each couple (climber, landscape) on UBQP landscapes. The average number of evaluations is reported below.

Landscape	F	B	W	W_2	W_4	W_8	W_{16}
$UBQP_{128,25}$	23398	23388	**23519**	23510	23511	23512	23517
	1k	7k	61k	2k	5k	12k	26k
$UBQP_{128,50}$	32784	32458	32913	32982	33052	**33063**	33006
	1k	8k	110k	2k	6k	16k	38k
$UBQP_{128,75}$	45435	45282	45571	**45610**	45600	45527	45524
	1k	7k	138k	2k	7k	16k	41k
$UBQP_{128,100}$	50104	49896	50475	50359	50592	50495	**50608**
	1k	8k	176k	2k	7k	17k	47k
$UBQP_{256,25}$	69183	69034	69518	69340	69503	**69574**	69505
	2k	31k	510k	6k	16k	37k	85k
$UBQP_{256,50}$	102037	101866	102121	102143	102168	102132	**102180**
	2k	34k	885k	7k	17k	43k	106k
$UBQP_{256,75}$	132076	131494	**133027**	132710	132791	132958	132829
	3k	35k	1254k	7k	19k	48k	119k
$UBQP_{256,100}$	144186	143724	144817	144597	144673	144671	**144826**
	2k	32k	1507k	7k	19k	51k	133k
$UBQP_{512,10}$	132133	131881	131954	**132232**	132202	132119	131902
	6k	136k	1614k	15k	34k	77k	170k
$UBQP_{512,50}$	284049	283560	**286723**	285842	286223	286696	286530
	8k	140k	8144k	21k	48k	122k	311k
$UBQP_{512,75}$	361205	360518	**362802**	361983	362528	362678	362713
	8k	138k	11020k	20k	50k	130k	338k
$UBQP_{512,100}$	419889	418458	421017	420398	421004	**421391**	421141
	7k	137k	12797k	20k	53k	133k	352k
$UBQP_{1024,10}$	350574	349469	351940	350826	351678	351928	**352137**
	17k	536k	15087k	39k	93k	220k	524k
$UBQP_{1024,50}$	797468	795302	**800153**	798871	799807	799851	799907
	22k	558k	72777k	53k	140k	354k	901k
$UBQP_{1024,75}$	998493	995156	1001466	999585	1000837	**1001775**	1001059
	22k	564k	95316k	54k	141k	366k	949k
$UBQP_{1024,100}$	1130314	1127151	1134202	1133432	1134057	1134332	**1134953**
	22k	568k	114228k	63k	147k	384k	1029k
$UBQP_{2048,10}$	991463	988944	**994765**	993698	994032	994163	994586
	49k	2230k	134784k	112k	264k	604k	1473k
$UBQP_{2048,25}$	1626645	1623603	1630519	1628570	1629699	1629879	**1630585**
	55k	2276k	325672k	126k	302k	734k	1886k
$UBQP_{2048,50}$	2377355	2373279	**2382892**	2380144	2381361	2381433	2381383
	60k	2295k	584078k	139k	345k	888k	2235k
$UBQP_{2048,100}$	3058752	3045996	**3074422**	3062042	3065725	3068646	3068596
	71k	2291k	951650k	168k	420k	1104k	2905k
$UBQP_{4096,10}$	2775627	2771590	2788231	2781304	2785430	2787136	**2788788**
	147k	9333k	1179075k	312k	713k	1671k	4089k
$UBQP_{4096,25}$	4552524	4545854	**4568967**	4558515	4562055	4567451	4568374
	160k	9364k	27611448k	350k	836k	2098k	5078k
$UBQP_{4096,50}$	6470526	6457298	**6495855**	6479307	6483570	6489013	6489455
	178k	9504k	4847505k	426k	1025k	2580k	6012k
$UBQP_{4096,100}$	9014549	8997240	**9053431**	9027282	9039464	9046036	9047765
	188k	9569k	7562717k	437k	1140k	2865k	7336k

Let us notice that we do not consider best improvement in this study since it requires the evaluation of the whole neighborhood at each step of the search and is regularly less efficient than other rules on considered landscapes.

Here, we investigate the efficiency of Iterated Local Search (ILS) [12] using F and W_k as pivoting rules dedicated to intensification phases. The considered ILS consists of iterating climbing processes until a maximal number of evaluations is reached. The first climbing process starts from a randomly generated solution, whereas the following ones start from a solution obtained by applying P random moves from the last obtained local optimum. The next section presents empirical results of ILS that only differs with the hill-climbing pivoting rule on several landscapes. We also analyze their behavior thanks to landscape features.

4   Experimental Analysis

4.1   Experimental Protocol

This experimentation is conducted on the set of 52 binary-string based fitness landscapes described in Sect. 2.2, using different fitness functions (NK, UBQP), sizes and ruggedness parameterization. It consists to compare the performances of iterated local search (which alternate hill-climbing and random perturbations) which only differ by the pivoting rule used during the climbing phases.

To obtain a fair comparison, all ILS runs are performed using different numbers of perturbations P on each landscape: $P \in \{5, 10, 15, 20\}$ on NK landscapes (higher values of P are useless here), and $P \in \{5, 10, 20, 30, 40, 50, 60, 70, 80\}$ on UBQP landscapes. Compared climbers are F, W_2, W_4, W_8, W_{16}. These different values of k allow various tradeoffs between quality of approximations and computation cost of climbers W_k. Recall that F is equivalent to W_1. Using $k = 2$ leads to the least precise approximation of worst improvement but is sufficient to induce a behavior quite different from first improvement F. Using $k = 16$ allow a very precise approximation of worst improvement and very similar results despite a reduced cost.

In the following we call ILS_F the iterated local search which uses a first improvement climbing rule and ILS_{W_k} iterated local search variants which use W_k. For each triplet (landscape, method, P) 100 runs starting from the same set of 100 randomly generated solutions are performed. The set of methods is composed of ILS_F, ILS_{W_2}, ILS_{W_4}, ILS_{W_8}, $ILS_{W_{16}}$. Each run stops after 100 million of evaluation and returns the fitness value of the best solution encountered during the search. For each triplet we record the average of the 100 resulting fitness values. In the next section we mainly report the best average fitness values obtained for each couple (landscape, method) as well as the value of P leading to this average.

4.2   Results

On considered NK landscapes (Table 3), we note that a first improvement based ILS always lead toward better local optima averages than using a worst improvement approximation W_k for climbing processes.

Table 3. Best average fitnesses obtained from 100 ILS runs on NK landscapes. Values between brackets correspond to the number of perturbations P leading to the reported best averages.

Landscape	ILS_F	ILS_{W_2}	ILS_{W_4}	ILS_{W_8}	$ILS_{W_{16}}$
$NK_{128,1}$	**.7245**[20]	**.7245**[20]	**.7245**[20]	.7166[20]	.7155[20]
$NK_{128,2}$	**.7423**[15]	.7415[15]	.7403[15]	.7394[15]	.7387[15]
$NK_{128,4}$	**.7958**[5]	.7956[5]	.7946[5]	.7941[5]	.7938[5]
$NK_{128,6}$	**.7995**[5]	.7955[5]	.7916[5]	.7892[5]	.7877[5]
$NK_{128,8}$	**.7949**[5]	.7896[5]	.7842[5]	.7818[5]	.7795[5]
$NK_{128,10}$	**.7847**[5]	.7784[5]	.7742[5]	.7717[5]	.7700[5]
$NK_{128,12}$	**.7724**[5]	.7676[5]	.7631[5]	.7601[5]	.7588[5]
$NK_{256,1}$	**.7200**[20]	.7179[20]	.7148[20]	.7124[20]	.7111[20]
$NK_{256,2}$	**.7425**[5]	.7396[10]	.7373[10]	.7351[10]	.7339[15]
$NK_{256,4}$	**.7917**[5]	.7899[5]	.7883[5]	.7876[5]	.7874[5]
$NK_{256,6}$	**.8007**[5]	.7957[5]	.7922[5]	.7907[5]	.7899[5]
$NK_{256,8}$	**.7892**[5]	.7828[5]	.7785[5]	.7751[5]	.7742[5]
$NK_{256,10}$	**.7782**[5]	.7721[5]	.7666[5]	.7642[5]	.7629[5]
$NK_{256,12}$	**.7663**[5]	.7602[5]	.7556[5]	.7527[5]	.7520[5]
$NK_{512,1}$	**.7040**[20]	.6984[20]	.6926[20]	.6894[20]	.6882[20]
$NK_{512,2}$	**.7453**[5]	.7419[10]	.7393[10]	.7381[10]	.7371[5]
$NK_{512,4}$	**.7806**[5]	.7770[5]	.7750[5]	.7742[5]	.7736[5]
$NK_{512,6}$	**.7940**[5]	.7899[5]	.7872[5]	.7858[5]	.7850[5]
$NK_{512,8}$	**.7886**[5]	.7842[5]	.7801[5]	.7783[5]	.7773[5]
$NK_{512,10}$	**.7781**[5]	.7731[5]	.7676[5]	.7651[5]	.7641[5]
$NK_{512,12}$	**.7671**[5]	.7612[5]	.7552[5]	.7524[5]	.7512[5]
$NK_{1024,1}$	**.7087**[15]	.7012[20]	.6942[20]	.6907[20]	.6898[20]
$NK_{1024,2}$	**.7428**[20]	.7385[20]	.7353[20]	.7332[15]	.7321[15]
$NK_{1024,4}$	**.7797**[5]	.7759[5]	.7736[5]	.7726[5]	.7718[5]
$NK_{1024,6}$	**.7890**[5]	.7857[5]	.7835[5]	.7817[5]	.7813[5]
$NK_{1024,8}$	**.7850**[5]	.7826[5]	.7801[5]	.7787[5]	.7787[5]
$NK_{1024,10}$	**.7753**[10]	.7735[5]	.7710[5]	.7694[5]	.7690[5]
$NK_{1024,12}$	**.7656**[5]	.7640[5]	.7609[5]	.7583[5]	.7582[5]

The better quality in average of local optima obtained through climbers using W_k is not sufficient to counterbalance the higher number of local optima reached with climbers F, since an ILS process only returns the best local optimum encountered. Note that the number of perturbations required to reach the best averages is always higher on smooth NK landscapes ($K = 1$ or $K = 2$). It could probably come from the fact that on smooth landscapes local optima are less numerous with larger basins of attraction. Successfully escaping from these local optima requires more random moves.

On UBQP landscapes (see Table 4), we observe that on smaller landscapes ($N \leq 1024$), almost all considered methods always reach the same local optimum, which we expect to be the global optimum[2].

Table 4. Comparative ILS results on UBQP landscapes (gaps from best averages).

Landscape	Best avg	ILS$_F$	ILS$_{W_2}$	ILS$_{W_4}$	ILS$_{W_8}$	ILS$_{W_{16}}$
UBQP$_{128,25}$	24087.0	$0.0_{[5-80]}$	$0.0_{[5-80]}$	$0.0_{[5-80]}$	$0.0_{[5-80]}$	$0.0_{[5-80]}$
UBQP$_{128,50}$	33440.0	$0.0_{[5-80]}$	$0.0_{[5-80]}$	$0.0_{[5-80]}$	$0.0_{[5-80]}$	$0.0_{[5-80]}$
UBQP$_{128,75}$	46180.0	$0.0_{[5-80]}$	$0.0_{[5-80]}$	$0.0_{[5-80]}$	$0.0_{[5-80]}$	$0.0_{[5-80]}$
UBQP$_{128,100}$	51130.0	$0.0_{[5-80]}$	$0.0_{[5-80]}$	$0.0_{[5-80]}$	$0.0_{[5-80]}$	$0.0_{[5-80]}$
UBQP$_{256,25}$	70861.0	$0.0_{[5-80]}$	$0.0_{[5-80]}$	$0.0_{[5-80]}$	$0.0_{[5-30.50]}$	$0.0_{[5-20]}$
UBQP$_{256,50}$	102914.0	$0.0_{[5-80]}$	$0.0_{[5-80]}$	$0.0_{[5-80]}$	$0.0_{[5-80]}$	$0.0_{[5-30]}$
UBQP$_{256,75}$	133641.0	$0.0_{[10-80]}$	$0.0_{[5-80]}$	$0.0_{[5-80]}$	$0.0_{[5-80]}$	$0.0_{[5-80]}$
UBQP$_{256,100}$	146377.0	$0.0_{[5-80]}$	$0.0_{[5-80]}$	$0.0_{[5-80]}$	$0.0_{[5-80]}$	$0.0_{[5-80]}$
UBQP$_{512,10}$	134112.0	$0.0_{[5-80]}$	$0.0_{[5-80]}$	$0.0_{[5-50]}$	$0.0_{[5-30]}$	$0.0_{[5-20]}$
UBQP$_{512,50}$	102914.0	$0.0_{[5-80]}$	$0.0_{[5-80]}$	$0.0_{[5-80]}$	$0.0_{[5-80]}$	$0.0_{[5-80]}$
UBQP$_{512,75}$	133641.0	$0.0_{[10-80]}$	$0.0_{[5-80]}$	$0.0_{[5-80]}$	$0.0_{[5-80]}$	$0.0_{[5-20]}$
UBQP$_{512,100}$	424728.0	$0.0_{[10-80]}$	$0.0_{[5-80]}$	$0.0_{[5-80]}$	$0.0_{[5-80]}$	$0.0_{[5-60]}$
UBQP$_{1024,10}$	356679.0	$0.0_{[60-80]}$	$0.0_{[40-50]}$	$-0.7_{[30]}$	$-0.7_{[20]}$	$-21.8_{[20]}$
UBQP$_{1024,50}$	808138.0	$0.0_{[50-80]}$	$0.0_{[40]}$	$0.0_{[20]}$	$-0.2_{[10]}$	$-16.6_{[10]}$
UBQP$_{1024,75}$	1007737.0	$0.0_{[30-80]}$	$0.0_{[20-40]}$	$0.0_{[10]}$	$-0.5_{[5]}$	$-8.3_{[10]}$
UBQP$_{1024,100}$	1145975.0	$0.0_{[20-80]}$	$0.0_{[20-80]}$	$0.0_{[20-80]}$	$0.0_{[10-40]}$	$0.0_{[10-20]}$
UBQP$_{2048,10}$	1004593.1	$-34.5_{[80]}$	$0.0_{[60]}$	$-64.2_{[40]}$	$-154.8_{[30]}$	$-390.0_{[30]}$
UBQP$_{2048,25}$	1641377.7	$-75.8_{[70]}$	$0.0_{[50]}$	$-117.6_{[20]}$	$-209.5_{[10]}$	$-197.5_{[10]}$
UBQP$_{2048,50}$	2398651.0	$-49.4_{[80]}$	$0.0_{[50]}$	$-51.7_{[30]}$	$-145.5_{[20]}$	$-597.8_{[10]}$
UBQP$_{2048,100}$	3100346.3	$-832.8_{[80]}$	$0.0_{[80]}$	$-382.5_{[40]}$	$-813.3_{[30]}$	$-1499.8_{[30]}$
UBQP$_{4096,10}$	2809642.0	$-953.7_{[80]}$	$0.0_{[80]}$	$-277.5_{[50]}$	$-818.1_{[30]}$	$-1033.7_{[30]}$
UBQP$_{4096,25}$	4597565.2	$-1536.5_{[80]}$	$0.0_{[80]}$	$-155.1_{[60]}$	$-731.0_{[30]}$	$-1640.4_{[20]}$
UBQP$_{4096,50}$	6530759.4	$-2914.8_{[80]}$	$0.0_{[80]}$	$-216.3_{[70]}$	$-1634.4_{[40]}$	$-3725.8_{[30]}$
UBQP$_{4096,100}$	9095931.9	$-4640.3_{[80]}$	$0.0_{[80]}$	$-49.3_{[60]}$	$-1940.0_{[30]}$	$-5023.4_{[30]}$

Oppositely, 1024-dimensional random NK landscapes are considered as large landscapes since even efficient evolutionary techniques fail to easily detect a global optimum. We then perform an additional analysis in order to obtain the

---

[2] We use the term *expected global optimum* when a same fitness is always reached by a set of methods. Constantly obtaining the same final solution (or fitness) does not guarantee its optimality, which could only be proved using complete methods.

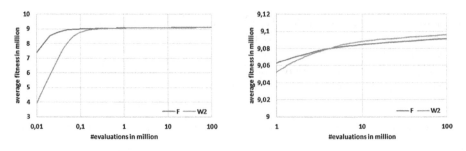

Fig. 2. Evolution of the best average fitness among all considered ILS_F and all considered ILS_{W_2} on UBQP landscape where $N = 4096$ and $d = 100$.

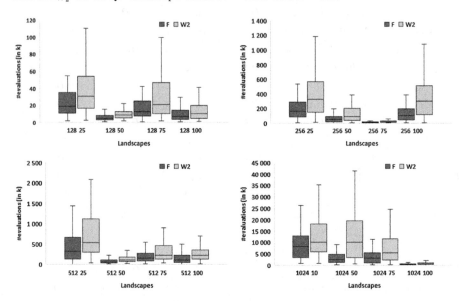

Fig. 3. Number of evaluations required to reach an expected global optimum from 100 runs of ILS_F and ILS_{W_2} on UBQP landscapes.

number of evaluations required by each ILS to reach an expected global optimum (on these landscapes derived from *easy* UBQP instances). Results are reported in Fig. 3 which compares, for each landscape, the computational budget used by F and W_2 for reaching an expected global optimum: ILS_F is faster than ILS_{W_2}, but the difference is relatively low.

On all larger UBQP landscapes $N \geq 2048$, ILS_{W_2} leads on average toward better local optima than ILS_F. Moreover, on 4096-dimensional UBQP landscapes, generating up to 4 and sometimes 8 improving neighbors before selecting the one with the lowest fitness can be more relevant than using first improvement.

Figure 2 depicts the evolution of the best average fitness during runs on an UBQP landscape instance ($N = 4096$, $d = 100$), for ILS_F and ILS_{W_2}. On this landscape, ILS_{W_2} starts to outperform ILS_F around 10 million evaluations. Let us notice that ILS_F and ILS_{W_2} have a similar compared evolution during the search on all considered large landscapes. Let us recall that single climbing processes using W_2 require more than twice as many evaluations than a climbing process using first improvement (see Table 2). In particular, on considered UBQP landscapes with $N \geq 2048$ a single first improvement climber requires on average 49k to 188k evaluations before terminating, whereas a W_2 climber requires from 112k to 437k evaluations.

Perturbations used by these iterated local search variants to escape from local optima (strategy: P random moves from the last local optimum reached) generally lead toward solutions slightly better than while restarting the search from a new random solution, especially since climbs following perturbations are faster than the first one. For instance, with the computational budget of the initial climbing process of ILS_{W_2}, ILS_F can proceed to many hill-climbings (as F is quicker than W_2, and following ILS_F climbings are also quicker than F). Thus the fact that it requires almost 10 million of evaluations for ILS_{W_2} to outperform ILS_F makes sense.

4.3   ILS Performance and Landscape Features

Experiments show a difference of overall efficiency between ILS_{W_k} and ILS_F on NK and UBQP landscapes. This subsection is devoted to obtaining insights on when using W_k within ILS leads toward better local optima. To this aim we measure for all considered landscapes the k-ruggedness evolution as well as the repartition of local optima.

k-ruggedness values are computed from sets of 100,000 pairs of solutions which are distant by k bits. In Tables 5 and 6, we report percentages corresponding to the ratios between the numbers of distinct 1-flips required to reach a given k-ruggedness value, and the maximal value $N - 1$. We also estimate the 1-ruggedness of landscapes (i.e. the proportion of 1-sign-epistasis) since local sign epistasis is an obstacle for climbers and directly affects their performance.

Results show that on NK landscapes (Table 5), the k-ruggedness grows faster on more rugged landscapes (when K increases) and is stable when N increases. For a given value of N, the 1-ruggedness is higher on more rugged landscapes and for a given value of K, the 1-ruggedness decreases on larger landscapes. This last observation is coherent since a single mutation has more impact on smaller NK landscapes than on larger ones. In random NK landscapes, the ruggedness is directly correlated with the parameters ratio K/N. So we use these observations as references to understand the structure of UBQP landscapes.

1-ruggedness values of UBQP landscapes (see Table 6) seem to increase according to their density parameter. This value is the only one, among considered values, differing on UBQP landscapes and corresponds to the 1-ruggedness of non-smooth NK landscapes of same size. Moreover, on larger landscapes, the 1-ruggedness decreases less than on NK landscapes of same size. The k-ruggedness evolution is similar for all UBQP landscapes, but does not correspond to the k-ruggedness evolution of any NK landscape. A k-ruggedness value greater than

Table 5. Ruggedness information of NK landscapes.

Landscape	k-ruggedness		1-rug.	Landscape	k-ruggedness		1-rug.
	$\geq$0.1	$\geq$0.25			$\geq$0.1	$\geq$0.25	
$NK_{128,1}$	15.7%	43.3%	0.5%	$NK_{256,1}$	16.1%	43.9%	0.3%
$NK_{128,2}$	7.9%	22.9%	1.2%	$NK_{256,2}$	7.5%	22.4%	0.6%
$NK_{128,4}$	3.1%	10.2%	3.1%	$NK_{256,4}$	3.1%	10.2%	1.5%
$NK_{128,6}$	2.4%	7.1%	5.2%	$NK_{256,6}$	1.9%	6.7%	2.6%
$NK_{128,8}$	1.6%	4.7%	7.4%	$NK_{256,8}$	1.2%	4.7%	4.1%
$NK_{128,10}$	1.6%	3.9%	9.6%	$NK_{256,10}$	1.2%	3.9%	5.3%
$NK_{128,12}$	0.8%	3.1%	11.8%	$NK_{256,12}$	0.8%	3.1%	6.5%
$NK_{512,1}$	16.7%	45.0%	0.1%	$NK_{1024,1}$	15.7%	42.7%	0.1%
$NK_{512,2}$	7.0%	22.3%	0.3%	$NK_{1024,2}$	7.6%	23.6%	0.2%
$NK_{512,4}$	2.9%	10.4%	0.8%	$NK_{1024,4}$	2.9%	10.1%	0.4%
$NK_{512,6}$	1.8%	6.4%	1.4%	$NK_{1024,6}$	1.8%	6.5%	0.7%
$NK_{512,8}$	1.2%	4.7%	2.1%	$NK_{1024,8}$	1.2%	4.8%	1.1%
$NK_{512,10}$	0.9%	3.7%	2.7%	$NK_{1024,10}$	0.9%	3.7%	1.4%
$NK_{512,12}$	0.7%	3.1%	3.6%	$NK_{1024,12}$	0.7%	3.1%	1.8%

Table 6. Ruggedness information of UBQP landscapes.

Landscape	k-ruggedness		1-rug.	Landscape	k-ruggedness		1-rug.
	$\geq$0.1	$\geq$0.25			$\geq$0.1	$\geq$0.25	
$UBQP_{128,25}$	5.5%	27.6%	2.3%	$UBQP_{256,25}$	5.1%	28.6%	1.5%
$UBQP_{128,50}$	4.7%	27.6%	3.1%	$UBQP_{256,50}$	5.5%	29.8%	2.1%
$UBQP_{128,75}$	5.5%	31.5%	3.6%	$UBQP_{256,75}$	5.5%	31.8%	2.7%
$UBQP_{128,100}$	5.5%	30.7%	4.3%	$UBQP_{256,100}$	5.1%	29.0%	2.8%
$UBQP_{512,10}$	5.9%	32.5%	0.6%	$UBQP_{1024,10}$	5.2%	30.5%	0.5%
$UBQP_{512,50}$	5.1%	32.3%	1.5%	$UBQP_{1024,25}$	4.9%	28.0%	1.0%
$UBQP_{512,75}$	4.9%	29.3%	1.8%	$UBQP_{1024,50}$	5.0%	28.6%	1.1%
$UBQP_{512,100}$	5.5%	32.5%	2.3%	$UBQP_{1024,100}$	5.2%	28.0%	1.5%
$UBQP_{2048,10}$	5.2 %	29.5%	0.3%	$UBQP_{4096,10}$	4.7%	28.8%	0.2%
$UBQP_{2048,25}$	5.0%	27.6%	0.7%	$UBQP_{4096,25}$	5.0%	28.9%	0.5%
$UBQP_{2048,50}$	5.0%	29.2%	0.9%	$UBQP_{4096,50}$	4.7%	28.5%	0.3%
$UBQP_{2048,100}$	4.8%	27.7%	1.0%	$UBQP_{4096,100}$	5.7%	30.0%	0.7%

0.1 is quickly reached (with few mutations), such as on medium-rugged NK landscapes. Yet, a k-ruggedness value superior to 0.25 is lately reached, such as on smooth NK landscapes. It means that the sign epistasis repartition, which directly affects the ruggedness, is different from the one of NK landscapes: the evolution of 1-ruggedness shows that the local sign epistasis is high on UBQP landscapes whereas the k-ruggedness global evolution leads to think that they globally have a reduced sign epistasis like smooth landscapes. Consequently, we assume that considered UBQP landscapes are locally rugged but globally smooth. Indeed, the 1-ruggedness values, which refer to the sign epistasis between neighboring solutions, reveal locally rugged landscapes. On the contrary, the k-sign-epistasis with high values of k, which refers to the sign epistasis between distant solutions, reveals globally smooth landscapes.

In addition we perform a sampling of local optima in considered landscapes. Local optima are collected from 1000 hill-climbing algorithms using a first improvement hill-climbing. In Tables 7 and 8 we report the average hamming distance between all distinct local optima found on each landscape as well as their numbers. Note that median distances are similar to average distances and are not reported.

Table 7. Average distance (d_{avg}) of distinct local optima on NK landscapes obtained from 1k first improvement climbers. #LO denotes the number of distinct local optima.

Landscape	d_{avg}	#LO	Landscape	d_{avg}	#LO
$NK_{128,1}$	24.48	991	$NK_{256,1}$	50.22	1000
$NK_{128,2}$	44.94	1000	$NK_{256,2}$	93.73	1000
$NK_{128,4}$	60.99	1000	$NK_{256,4}$	121.43	1000
$NK_{128,6}$	63.55	1000	$NK_{256,6}$	126.66	1000
$NK_{128,8}$	63.86	1000	$NK_{256,8}$	127.69	1000
$NK_{128,10}$	63.97	1000	$NK_{256,10}$	127.92	1000
$NK_{128,12}$	63.97	1000	$NK_{256,12}$	127.99	1000
$NK_{512,1}$	101.74	1000	$NK_{1024,1}$	200.02	1000
$NK_{512,2}$	187.37	1000	$NK_{1024,2}$	366.87	1000
$NK_{512,4}$	243.29	1000	$NK_{1024,4}$	486.74	1000
$NK_{512,6}$	253.28	1000	$NK_{1024,6}$	507.17	1000
$NK_{512,8}$	255.40	1000	$NK_{1024,8}$	510.56	1000
$NK_{512,10}$	255.85	1000	$NK_{1024,10}$	511.75	1000
$NK_{512,12}$	255.96	1000	$NK_{1024,12}$	511.95	1000

Results on NK landscapes (Table 7) show that on sufficiently rugged landscapes, the average hamming distance is close to $N/2$. It means that local optima are uniformly distributed in the landscape, since $N/2$ corresponds to the average hamming distance between two random solutions. On smooth NK landscapes the average hamming distance between considered local optima is lower, meaning that local optima are packed in a relatively reduced area of the landscape.

On UBQP landscapes (Table 8), the average distances of local optima are largely lower than $N/2$ and even often lower than distances observed on very smooth NK landscapes. Moreover, the number of distinct local optima found on smaller considered landscapes is much lower than on same size NK landscapes and could explain why associated problem instances are easier to solve. The observed number and repartition of local optima on UBQP landscapes tend to confirm the assumed structure of such landscapes, that is locally rugged but globally very smooth.

Table 8. Average distance of distinct local optima on UBQP landscapes.

Landscape	d_{avg}	#LO	Landscape	d_{avg}	#LO
$UBQP_{128,25}$	26.26	815	$UBQP_{256,25}$	52.18	990
$UBQP_{128,50}$	24.20	378	$UBQP_{256,50}$	26.45	640
$UBQP_{128,75}$	21.5	488	$UBQP_{256,75}$	30.84	566
$UBQP_{128,100}$	23.58	468	$UBQP_{256,100}$	41.98	908
$UBQP_{512,10}$	59.40	999	$UBQP_{1024,10}$	156.11	1000
$UBQP_{512,50}$	63.52	976	$UBQP_{1024,50}$	157.62	1000
$UBQP_{512,75}$	59.27	980	$UBQP_{1024,75}$	130.15	1000
$UBQP_{512,100}$	59.94	969	$UBQP_{1024,100}$	133.19	1000
$UBQP_{2048,10}$	269.67	1000	$UBQP_{4096,10}$	536.15	1000
$UBQP_{2048,25}$	230.67	1000	$UBQP_{4096,25}$	512.03	1000
$UBQP_{2048,50}$	238.1	1000	$UBQP_{4096,50}$	483.50	1000
$UBQP_{2048,100}$	307.22	1000	$UBQP_{4096,100}$	486.59	1000

The fact that ILS_{W_k} is more efficient on UBQP landscapes than on NK landscapes can be explained by the structure of these landscapes. UBQP landscapes are globally smooth, which at first glance is not an advantage for climbing strategies based on worst improvement, yet such landscapes are locally rugged. Pivoting rules based on worst improvement help to bypass low quality local optima and avoid to be trapped by them. Such strategies seem to lead toward higher local optima on UBQP landscapes than on NK landscapes compared to a basic first improvement.

On NK landscapes, the quality of local optima obtained during ILS_{W_k} runs is not sufficiently improved compare to those obtained during an ILS_F, which is able to perform more climbing processes with the same number of evaluations. On UBQP landscapes, ILS_{W_k} achieves a better balance between local optima quality and number of evaluations to reach them than ILS_F.

5   Conclusion

Most climbing processes of metaheuristics are based on first or best improvement pivoting rules. As stated by a previous study [2], worst improvement is able to reach good local optima despite its counter-intuitive nature. In this work, we integrated worst improvement in an Iterative Local Search context in order to obtain insights on the potential interest of using such a pivoting rule within advanced metaheuristics. We performed experiments on binary-string problems (NK functions and UBQP) and combined results observations with fitness landscape analysis. In these experiments we confronted worst improvement based ILS with first improvement based ILS. In order to obtain tradeoffs between local optima quality and number of evaluations, we included intermediate variants which use different levels of approximation of worst improvement. Experiments show that on NK landscapes, first improvement remains the most adequate pivoting rule in an ILS context, thanks to its small number of evaluations that allow it to reach more local optima. However, on large UBQP landscapes, worst improvement approximations are able to outperform first improvement in an ILS context, despite the heavier computation cost of each climbing process.

Thanks to landscape features (ruggedness information, distribution of local optima), we deduced that the structure of UBQP landscapes seems to be globally smooth but locally rugged and that local optima are more packed within the search space. These facts induce an advantage of the approximated worst improvement climbing strategy which has a tendency to avoid to be prematurely trapped in low quality local optima. On such landscapes this advantage has more impact than the small number of evaluations induced by first improvement climbers and allows iterated searches to be efficient.

It should be interesting to extend this type of work to other problems and/or other metaheuristics in order to determine if alternative pivoting rules could be successfully used in other contexts.

References

1. Sörensen, K.: Metaheuristics—the metaphor exposed. Int. Trans. Oper. Res. **22**(1), 3–18 (2015)
2. Whitley, D., Howe, A.E., Hains, D.: Greedy or not? Best improving versus first improving stochastic local search for MAXSAT. In: AAAI Conference on Artificial Intelligence (2013)
3. Wright, S.: The roles of mutation, inbreeding, crossbreeding, and selection in evolution. vol. 1 (1932)
4. Ochoa, G., Tomassini, M., Verel, S., Darabos, C.: A study of NK landscapes' basins and local optima networks. In: Conference on Genetic and Evolutionary Computation, pp. 555–562. ACM (2008)
5. Basseur, M., Goëffon, A.: Climbing combinatorial fitness landscapes. Appl. Soft Comput. **30**, 688–704 (2015)
6. Malan, K.M., Engelbrecht, A.P.: A survey of techniques for characterising fitness landscapes and some possible ways forward. Inf. Sci. **241**, 148–163 (2013)
7. Kauffman, S.A., Weinberger, E.D.: The NK model of rugged fitness landscapes and its application to maturation of the immune response. J. Theoret. Biol. **141**(2), 211–245 (1989)
8. Gary, M.R., Johnson, D.S.: Computers and intractability: a guide to the theory of NP-completeness (1979)

9. Basseur, M., Goëffon, A., Lardeux, F., Saubion, F., Vigneron, V.: On the attainability of NK landscapes global optima. In: 7th Annual Symposium on Combinatorial Search (2014)

10. Ochoa, G., Verel, S., Tomassini, M.: First-improvement vs. best-improvement local optima networks of NK landscapes. In: Schaefer, R., Cotta, C., Kołodziej, J., Rudolph, G. (eds.) PPSN 2010. LNCS, vol. 6238, pp. 104–113. Springer, Heidelberg (2010). https://doi.org/10.1007/978-3-642-15844-5_11

11. Basseur, M., Goëffon, A.: Hill-climbing strategies on various landscapes: an empirical comparison. In: Genetic and Evolutionary Computation Conference (GECCO), pp. 479–486. ACM (2013)

12. Loureno, H.R., Martin, O.C., Stutzle, T.: Iterated local search. Int. Ser. Oper. Res. Manag. Sci. 321–354 (2003)

Automatic Grammar-Based Design of Heuristic Algorithms for Unconstrained Binary Quadratic Programming

Marcelo de Souza[1,2]([✉]) and Marcus Ritt[2]

[1] Santa Catarina State University, Ibirama, SC, Brazil
marcelo.desouza@udesc.br
[2] Federal University of Rio Grande do Sul, Porto Alegre, RS, Brazil
marcus.ritt@inf.ufrgs.br

Abstract. Automatic methods have been applied to find good heuristic algorithms to combinatorial optimization problems. These methods aim at reducing human efforts in the trial-and-error search for promising heuristic strategies. We propose a grammar-based approach to the automatic design of heuristics and apply it to binary quadratic programming. The grammar represents the search space of algorithms and parameter values. A solution is represented as a sequence of categorical choices, which encode the decisions taken in the grammar to generate a complete algorithm. We use an iterated F-race to evolve solutions and tune parameter values. Experiments show that our approach can find algorithms which perform better than or comparable to state-of-the-art methods, and can even find new best solutions for some instances of standard benchmark sets.

Keywords: Automatic algorithm configuration
Grammatical evolution · Metaheuristics

1 Introduction

Manually developing heuristic algorithms to solve optimization problems is laborious and often biased. Engineering an effective heuristic requires to define adequate algorithmic components (e.g. neighborhoods for local searches) and the selection of a meta-heuristic, i.e. a heuristic strategy that defines how the components operate to explore the solution space (e.g. a tabu search). Since heuristic methods are difficult to analyze, there is almost no guiding theory, and the main development step consists in implementing a heuristic method and evaluating it experimentally. This is repeated in order to improve the algorithms by introducing new problem-specific components and by tuning their parameters. The experimental evaluation can take a large amount of time, which is partially spent testing ineffective components and weak heuristic strategies. At the same time good strategies may be overlooked, since heuristic design space can be very large,

© Springer International Publishing AG, part of Springer Nature 2018
A. Liefooghe and M. López-Ibáñez (Eds.): EvoCOP 2018, LNCS 10782, pp. 67–84, 2018.
https://doi.org/10.1007/978-3-319-77449-7_5

and the heuristic performance can depend on a good parameter setting. As a consequence, the design space is often not explored systematically. Additionally, designers often use shortcuts, such as short test runs, or choose a meta-heuristic only based on previous experience (e.g. a genetic algorithm, which is one of the oldest and best explored meta-heuristics). This can lead to additional biases in the design.

In order to overcome these problems, automatic algorithm engineering methods or algorithm configurators have been developed. They explore the design space and choose the best heuristic strategy combining a given set of components and finding the best parameter values. Some examples with a variety of application studies are ParamILS [1], GGA [2], SMAC [3], and irace [4].

In this paper, we focus on finding good heuristics for unconstrained Binary Quadratic Programming (BQP). Given a matrix $Q = (q_{ij}) \in \mathbb{R}^{n \times n}$ BQP asks to

$$\text{maximize} \quad x^t Q x,$$
$$\text{subject to} \quad x \in \{0, 1\}^n.$$

Many problems can be reduced to binary quadratic optimization. For example, problems from machine scheduling [5], satisfiability problems [6], maximum clique problems [7], and others can be reduced to BQP. An example is the decision version of the maximum clique (MC) problem. Given an undirected graph $G = (V, E)$ and a value k, is there clique (i.e. a complete subgraph) of size k or more? With binary variables $x_v \in \{0, 1\}$, $v \in V$, and weights $q_{uv} = 1$ for $\{u, v\} \in E$, and $q_{uv} = -\binom{n}{2}$ otherwise, we have a "yes"-instance of MC iff BQP has a solution of value $\binom{k}{2}$ or more. This reduction can be done in polynomial time and since MC is NP-hard shows that BQP is NP-hard, too. BQP remains hard if there is unique solution [8]. Kochenberger et al. [9] present general reduction techniques to BQP and list more than twenty problems that can be reduced to BQP.

To find better heuristics for BQP, we first extract the main problem-specific components of state-of-the art algorithms from the literature. These components include constructive heuristics, neighborhoods for local searches, perturbation strategies for iterated local searches, and solution recombination strategies, such as path relinking. We then represent the design space of meta-heuristic strategies using these components, such as iterated local searches, by a context-free grammar. A concrete strategy is represented by a complete derivation in this grammar, and thus by a sequence of decisions which rules to apply. These decision are represented by a sequence of categorical choices. In this way, the task of the automatic configuration procedure is to find good sequences of choices and good values of the parameters of the heuristic components. We then use irace as a configurator in the combined choices and parameter space and evaluate the resulting heuristic strategies.

The main contributions of this paper are: (1) an experimental study of the automatic design of heuristics for BQP; (2) the extraction of specific heuristic components for BQP from state-of-the-art algorithms; and (3) a set of competitive heuristic algorithms automatically extracted from those components.

The rest of the paper is organized as follows. Section 2 discusses related work in automatic algorithm configuration and heuristic methods for BQP. Section 3 gives details about the components we have extracted from state-of-the-art algorithms and the grammar-based approach to automatically design heuristic algorithms for BQP. Section 4 presents computational experiments and discusses the algorithms found. Finally, Sect. 5 concludes and points to future work.

2   Related Work

There are different tools for automatic algorithm configuration. ParamILS [1] applies an iterated local search in the configuration space. The method starts from the best of a number of random configurations, and then repeatedly searches a local minimum and perturbs the current solution. The local search accepts the first improvement in a neighborhood which modifies a single parameter value. The perturbation consist in a number of random moves in the neighborhood. ParamILS uses adaptive strategies to reduce the cost for comparing configurations of stochastic heuristics. GGA [2] propose a gender-based genetic algorithm for parameter tuning. Individuals represent parameter settings, and the parameters are organized in decision trees, which can represent the coupling of parameters, which is taken into account for crossover. The candidates are divided into competitive and non-competitive genders. A more recent version, GGA++ [10] guides the search using a model to estimate the empirical performance of candidates. Sequential model-based algorithm configuration (SMAC) [3] also applies a model to estimate algorithm performance. This model is used to find the most promising candidates in parameter space to evaluate next. The authors also propose a distributed version called dSMAC [11]. The irace method [4] applies iteratively Friedman-races [12] to candidates from a set of elite configurations. We have selected irace for our experiments, since it is a flexible, state-of-the-art configurator which supports categorical, ordinal, and real parameters, as well as conditional parameters. The method will be explained in more detail below.

There are many studies of the application of configurators for designing algorithms for combinatorial optimization problems. KhudaBukhsh et al. [13] present the automatic design of stochastic local search (SLS) solvers for the propositional satisfiability (SAT) problem. They apply ParamILS to find which components of SLS to use, obtaining results that outperform state-of-the-art algorithms. López-Ibáñez and Stützle [14] apply automatic algorithm configuration to design multi-objective ant colony optimization (MOACO) algorithms. They use irace to tune MOACO components and show that automatically designed algorithms can outperform MOACO algorithms found in the literature. A similar approach is found in Bezerra et al. [15], who propose the automatic design of evolutionary algorithms for multi-objective optimization problems. In both cases, the automatic approach can combine components from different algorithms of the literature, producing new state-of-the-art approaches.

Several researchers propose to represent the heuristic and parameter space by grammars [16–18]. A configurator then searches the set of decisions that leads to the best derivation. A common strategy is to represent the choices made in the rules of the grammar by numerical values, called codons. However, several authors have pointed out that simple codon-based approaches have problems with redundancy and locality and have proposed to use rule-based tokens [19–22]. For recursive rules a limit on the number of expansions has to be defined beforehand. Mascia et al. [21,22] additionally make parameters that are not used in all derivations conditional. In this work, we follow the same approach.

BQP is NP-hard and current exact methods can only solve small instances (for more details about the exact methods see, e.g., [23]). For this reason most research focuses on heuristic methods to solve medium and large instances in the range of 2500 to about 15000 variables. Palubeckis [24] proposes an iterated tabu search. His approach iteratively perturbs the best found solution and applies a tabu search procedure. Glover et al. [25] also propose an iterated tabu search, but add a diversification strategy based on a set of elite solutions, which provides initial solutions to the search procedure. The perturbation is also based on the elite set, scoring variables according to the frequency in the elite solutions. Wang et al. [26] propose a populational heuristic that keeps an elite solution set and recombines them by two different path-relinking strategies. They also apply a tabu search to the generated solutions to improve them, updating the elite set when better solutions are found. The solution-specific components of these methods are used in our approach and will be explained in more detail in the next section.

3   Proposed Approach

Our approach is based on two main components: (1) a grammar that represents the space of heuristic strategies and their parameters, based on a given set of algorithmic components, such as constructive algorithms and neighborhoods; (2) a configurator that finds the best grammar instantiation. This section details these two components.

3.1   Grammar and the Heuristic Search Space

Algorithm configuration by grammatical evolution represents the algorithm design space by grammar. Figure 1 shows the grammar we propose to model the design space of BQP in Backus-Naur form. The grammar consists of a set of rules, through which the heuristic algorithms can be instantiated. Each rule describes a decision, i.e., a component to be chosen. The start symbol of the grammar is the non-terminal ALG in line 1. Starting from ALG, three main heuristic strategies can be chosen: heuristics based on local *search*, on solution *construction*, or on *recombination* of candidates from a population of solutions. In the following we describe all three strategies in more detail.

Search methods. The search-based heuristics (line 2) start with an initial solution and perform modifications on it, in order to explore the search space. The solutions obtained by the application of all possible modifications form a solution

1	<ALG> ::= <SEARCH> | <CONSTRUCTION> | <RECOMBINATION>
2	<SEARCH> ::= LS(<IMPROVEMENT>) | NMLS(<IMPROVEMENT>) | <TS>
	| ILS(<SEARCH>, <PERT>) | ILSE(<SEARCH>, <PERT>)
3	<IMPROVEMENT> ::= FI | FI-RR | BI | SI | SI-PARTIAL | SI-PARTIAL-RR
4	<TS> ::= STS | RTS
5	<PERT> ::= RANDOM(<STEP>) | LEAST-LOSS(<STEP>) | DIVERSITY(<STEP>)
6	<STEP> ::= UNIFORM | GAUSSIAN | EXPONENTIAL | GAMMAM
7	<CONSTRUCTION> ::= GRA(<CONSTRUCTOR>) | GRASP(<CONSTRUCTOR>, <SEARCH>)
8	<CONSTRUCTOR> ::= ZERO | HALF
9	<RECOMBINATION> ::= RER(<IMPROVEMENT>, <SEARCH>)

Fig. 1. The grammar that describes the heuristic search space

neighborhood. Then, search methods apply some strategy to select the next solution from the neighborhood (line 3). Given that in the BQP the neighborhood of a solution is the set of solutions with one modified variable (a flip in a position of the vector x), we access the neighbors in the order of the variables. The first improvement (FI) strategy selects the first neighbor that improves the solution. We can apply a round-robin strategy (FI-RR), starting the exploration from the position where the previous one has finished. The best improvement (BI) strategy selects a neighbor that improves the solution most. The some improvement (SI) strategy selects a random improving neighbor. A variant, called some improvement with partial exploration (SI-PARTIAL) considers only the first $f\%$ of variables in the exploration for some improvement. If no improving neighbor is found, the rest of variables is explored. Alternatively, we can use a round robin strategy to start the exploration from the position where the previous one has finished (SI-PARTIAL-RR).

The simplest search-based method is the local search (LS), which iteratively applies an improving modification until no better neighbor is found. In order to avoid local optimum, a common strategy is to select a random neighbor with probability p, and an improving neighbor with probability $1 - p$. This method is called non-monotone local search (NMLS). Tabu search (TS) was first proposed by Glover [27] and consists of a local search that keeps a list of prohibited solutions (called tabu list), in order to avoid the search coming back to previous visited solutions in a short-term period. When a solution is selected, the modified variable is stored in the tabu list for some number of iterations (called the tabu tenure). During this period, this variable cannot be changed. There are different strategies to define the tabu tenure, like a constant, or a value according to the instance size. The simple tabu search (STS) always apply a best improvement strategy to select a neighbor. A randomized tabu search (RTS) applies a random move with a probability p. Commonly used stopping criteria for tabu search are a maximum number of iterations or a maximum number of iterations in stagnation. For example, Palubeckis [24] proposes the maximum number of iterations 15000 if the instance has more than 5000 variables, 12000 if it has between 3000 and 5000 variables, and 10000 for instances up to 3000 variables.

Iterated local search (ILS) was proposed by Lourenço et al. [28] and iteratively applies a local search, followed by a perturbation step (PERT). When the search procedure ends, the current solution is at a local minimum and is perturbed in order to escape it. The iterated local search can also be combined with an elite strategy (ILSE). The elite set stores the best solutions found so far, which are used as initial solutions for a perturbation and a search step. We also introduce some perturbation strategies. The first one selects variables to flip at random (RANDOM). The second strategy ranks variables according to the loss when flipping its value. Then, the LEAST-LOSS approach randomly selects one of the b variables of least loss. When using an elite set, the DIVERSITY strategy ranks variables according to the frequency in the elite set.

Our grammar allows the combination of the iterated local search with any search and perturbation procedures. Therefore, we can instantiate state-of-the-art algorithms such as the one proposed by Palubeckis [24], which consists of an iterated local search that applies a tabu search and the least-loss perturbation procedure. We also can instantiate the diversification method of Glover et al. [25], selecting the iterated local search elite with a tabu search and a diversity perturbation procedure. In this case, the diversification-based perturbation proposed by Glover et al. scores variables using a parameter β as a weight factor for the frequency contribution, and then selects variables to be randomly assigned to 0 or 1. The probability of selection is proportional to the variable's score, using the parameter λ to define the importance of the score in this step.

The grammar also has several strategies to define the size of the perturbation (<STEP>), i.e., the number of variables that will be flipped. Given the instance size n, the GAMMAM strategy defines the perturbation size as n/g. For the other strategies, the parameters d_1 and d_2 defines the interval of possible perturbation sizes as $[d_1, n/d_2]$. The UNIFORM strategy selects a value of the interval according to an uniform distribution. The GAUSSIAN and EXPONENTIAL strategies apply a Gaussian and exponential distribution, respectively.

Construction methods. The construction-based methods are based on the heuristics of Merz and Freisleben [29]. They start with an empty solution and iteratively set values to its variables. The variable is chosen according to an α-greedy algorithm, which randomly selects one of the $\alpha\%$ best variables. The greedy randomized adaptive (GRA) algorithm repeatedly constructs m solutions and picks the best one. Another approach, proposed by Feo and Resende [30], is the greedy randomized adaptive search procedure (GRASP), which applies a search procedure whenever a solution is constructed. A first strategy for the construction process for BQP starts with $x = 0$, and then sets variables to one (component ZERO in line 8). An alternative strategy starts all variables at 0.5, and then sets the values to zero or one (HALF).

Recombination methods. The recombination-based method implements the idea of evolving a population of solutions. Wang et al. [26] propose the repeated elite recombination (RER) method, which stores an elite set with the best solutions. The elite set is initially filled with random solutions after applying the search procedure to them. Then, each pair of solutions from the elite set is recombined using path relinking. Next, a search procedure is applied to the best

Table 1. Method's parameters (n is the instance size).

Param.	Type	Method	Description	Values
t	cat	`<TS>`	Strategy for tabu tenure	$\{t_1, t_2, t_3, t_4, t_5\}$
t_v	int	`<TS>` (t_1)	Constant for tabu tenure	$[1, 50]$
t_p	int	`<TS>` (t_2)	Tabu tenure is $(t_p \times n)/100$	$[10, 80]$
t_d	int	`<TS>` $(t_3$ and $t_4)$	Tabu tenure is n/t_d	$[2, 500]$
t_c	int	`<TS>` (t_4)	Tabu tenure is $n/t_d + c \in [0, t_c]$	$[1, 100]$
s	cat	`<TS>`	Strategy for maximum stagnation	$\{s_1, s_2, s_3\}$
s_v	int	`<TS>` (s_1)	Constant for maximum stagnation	$[500, 100000]$
s_m	int	`<TS>` (s_2)	Max. stagnation is $s_m \times n$	$[1, 100]$
i	cat	`<TS>`	Strategy for maximum iterations	$\{i_1, i_2, i_3\}$
i_v	int	`<TS>` (i_1)	Constant for maximum iterations	$[1000, 50000]$
p	real	`NMLS; RTS`	Probability of a random move	$[0.0, 1.0]$
f	int	`SI-PARTIAL[-RR]`	Size of the partial exploration	$[5, 50]$
d_1	int	`<PERT>`	Min. perturbation size	$[1, 100]$
d_2	int	`<PERT>`	Max. perturbation size is n/d_2	$[1, 100]$
g	int	`GAMMAM`	Perturbation size is n/g	$[2, 100]$
b	int	`LEAST-LOSS`	Candidate variables for pert	$[1, 20]$
β	real	`DIVERSITY`	Frequency contribution	$[0.1, 0.9]$
λ	real	`DIVERSITY`	Selection importance factor	$[1.0, 3.0]$
r	int	`ILSE`	Elite set size for `ILSE`	$[1, 30]$
e	int	`RER`	Elite set size for `RER`	$[1, 20]$
γ	real	`RER`	Distance scale	$[0.1, 0.5]$
α	real	`<CONSTRUCTION>`	Greediness of the construction	$[0.0, 1.0]$
m	int	`<CONSTRUCTION>`	Number of repetitions	$[10, 100]$

solution found by path relinking. It replaces the worst solution in the elite set, if its quality is better. The recombination plus search step is performed repeatedly.

Given a pair of solutions, path relinking applies a local search to the first one, allowing only modifications that approximate it to the second solution. In other words, the modification of the variable i is only allowed if the new value is equal the value of variable i in the second solution. For this exploration, path relinking uses some improvement strategy. If using best improvement and no improving neighbor is found, the exploration selects the best neighbor (solution of the highest quality). When using the other strategies, if no improving neighbor is found, the exploration selects a random neighbor. Wang et al. [26] apply two different path relinking methods. The first one (PR1) uses the best improvement strategy, while the second (PR2) always selects a random neighbor. Moreover, PR1 and PR2 require that path relinking returns a solution with minimum and maximum distances from the endpoints as $d_{min} = \gamma \times H$ and $d_{max} = H - d_{min}$, where H is the Hamming distance between both solutions. If no such solution is found, the result is the starting solution s.

Parameters. Besides the algorithmic components, the grammar also defines the possible parameter values of each algorithm. For example, the constructive approaches require the definition of the number of repetitions. In this way, a complete derivation of the grammar not only gives an algorithm, but also its parameters values. Table 1 shows all parameters, their type, the related method, and possible values. Categorical parameters (like the different strategies to compute tabu tenure) have a set of limited possible values. For numerical parameters, we define the correspondent interval of possible values. Parameters t, s, and i choose the strategies for tabu tenure, maximum stagnation and maximum iterations, respectively. Strategies t_1 to t_4 are detailed in the table, and strategy t_5 is Palubeckis' rule for computing the tabu tenure. Strategy i_2 allows at most n iterations, while i_3 and s_3 allow iterations and stagnation ∞. The rest of parameters were explained above.

3.2   Automatic Design Using irace

We can see that our grammar allows to combine all the heuristic components, in order to: (1) generate the algorithms found in the literature of BQP; and (2) generate new and hybrid algorithms through the combination of these components. To perform the exploration of this search space, we use the irace tool described in [4]. irace implements an iterated racing procedure that keeps a set of elite candidates. Iteratively, it selects a candidate θ and an instance i, and then calls the target algorithm with both elements. The result metric $\varphi(\theta, i)$ is then returned to irace, which uses this value to rank different candidates. This process is repeated until irace has used the available budget of algorithm runs.

In the irace procedure, each parameter to be tuned has an associated probability distribution. Numerical and ordinal parameters have a truncated normal distribution, while categorical parameters have a discrete distribution. In each iteration, irace generates candidates according to the distribution

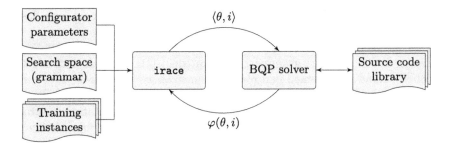

Fig. 2. General idea of the automatic algorithm configuration process

associated to their parameters, and then performs a racing to select the best candidates. Finally, the sampling distributions are updated for the next iteration, in order to bias the sampling towards the best candidates. For the selection phase, irace applies the non-parametric Friedman test to discard non-competitive configurations.

The general operation of the automatic process we use in this paper is depicted in Fig. 2. The grammar and a set of training instances are inputs to irace. We use the default values for the configurator parameters. The candidate is a complete heuristic method with the respective parameter values, and the target algorithm is the solver for the BQP. The source code library implements all heuristic components and is used by the BQP solver to run a candidate θ. We use the quality of the best found solution as result metric.

We can see that our grammar is recursive in the rule of line 2. However, unlimited depth is unhelpful and we limit the recursiveness by not allowing an iterated local search within another iterated local search. Thereby, we reduce the size of the search space defined by the grammar, but keep it possible to produce hybrid meta-heuristics, since it is based on fine-grained components and can flexibly combine them. Considering only the possible combination of components, our grammar can produce 3152 different algorithms. This number increases substantially if we consider the possible values of the 3 categorical and 15 integer parameters, and is infinite considering the 5 real parameters.

We deal with the grammar using a sequence of categorical choices representation. Each choice is an integer number, which defines the decision taken in the respective rule. For example, the rule

```
<ALG> ::= <SEARCH> | <CONSTRUCTION> | <RECOMBINATION>
```

has three possible values. A categorical choice that represents this rule will assume a value in the integer interval $[0, 2]$. Despite using integer values for each option, we follow the parametric representation presented by Mascia et al. [22] and implement these choices as categorical parameters in irace, to minimize the problem of locality. Two neighboring choice values (e.g. 0 and 1) not necessarily generate neighboring algorithms. In this example, changing the choice value from 0 to 1 will completely change the resulting algorithm, from a search-based to a construction-based method.

A second factor that leads to locality problems is using a single sequence of choices to decide over the grammar, as adopted in [31]. In each decision, the next choice value is consumed in a round-robin strategy. This means that a single modification in some choice value can change the rule related to all next choices. To avoid this, we adopt the idea of structured representation for grammatical evolution presented in [20]. Each rule has its own sequence of choices. Thereby, a specific choice corresponds to a specific rule, independent of the values of previous choices in the grammar derivation. Moreover, since each choice is associated to the same rule, the corresponding categorical variable can assume exactly the number of choices. In addition to reducing the search space, this avoids redundancy, i.e., two choice values that lead to the same algorithm.

We also limit the maximum number of choices for each rule. This is the maximum number of times that this rule can be used to generate an algorithm. Most rules have only one choice, since they can be applied at most once. Rule <MODIFICATION> has two choices, because two different modification strategies can be necessary in a single algorithm (e.g. choosing the recombination method with a local search). Defining only the minimal amount of choices is possible because the grammar is non-recursive. This feature reduces the search space size, making the task of automatic configuration easier.

Finally, some choices are conditional. The choice for rule <ALG> has to always have a value, because this decision is taken in all possible algorithms that the grammar can generate. This is not the case for the rest of rules. For example, rule <TS> is applied only if a tabu search is selected in rule <SEARCH>. In other words, the structure of categorical choices reflects a relationship between them, defining the conditions to the need of each specific choice. We implemented these conditions, such that the configuration task needs to set values only for required choices. This feature reduces even more the search space and also avoids redundancy.

4   Experiments and Results

This section presents our experiments and discusses the obtained results. We use three different instance sets. The first is a set of 10 instances proposed by Beasley. They have 2500 variables and 10% density, i.e., approximately 10% of the elements of matrix Q have non-zero values. The second is a set of 21 instances proposed by Palubeckis [24]. They have between 3000 and 7000 variables and densities from 50% to 100%. The coefficients of these instances are uniform random integers in the interval $[-100, 100]$. The last set contains 54 instances of the MaxCut problem[1]. They have between 800 and 2000 variables and densities less than 1%. The non-zero elements assume values in $\{-1, 1\}$.

All algorithm components were implemented in C++, using the GNU GCC compiler version 5.3.1. We ran all experiments on a Linux platform running

---

[1] Beasley and MaxCut instances can be found in http://biqmac.uni-klu.ac.at.

on a computer with an 8-core AMD FX-8150 CPU, with 3.6 GHz and 32 GB memory. Our machine is approximately four times faster than the machine used by Palubeckis [24], so we used 25% of the time limit used by Palubeckis. This results in a time limit of 150 s for the Beasley instances and 225 s, 450 s, 900 s, 1350 s, and 2250 s for instances of Palubeckis with 3000, 4000, 5000, 6000, and 7000 variables, respectively. For the MaxCut instances, we used 66.6% (1200 s) of the time limit used by Wang et al. [26], because our machine is not more than 33% faster than their machine.

4.1 Tuning with a Single Instance Set

The density and coefficients of the Beasley and Palubeckis instances, are significantly different from those of the MaxCut instances. Therefore, we test in our first experiment an individual tuning for these two instance groups. For each of the two experiments, we used samples of the instances to tune an algorithm with a budget of 2000 runs, and then validate it using the complete set of instances. The first tuning uses the instances p3000-1, p4000-1, p5000-1, p6000-1, and p7000-1 from Palubeckis, and the second uses MaxCut instances G1, G5, G10, G15, G20, G25, G30, G35, G40, G45, and G50. We call the algorithm trained on the Palubeckis instances AAC_P, and the one trained on the MaxCut instances AAC_M.

Table 2. Average absolute gap on the different instance sets.

Inst.	ITS	D^2TS	PR1	PR2	AAC_P	AAC_M
Beasley	2.0	**0.0**	1.34	5.84	**0.0**	34.0
Palubeckis	1109.6	2082.9	457.1	690.4	**186.8**	2424.3
MaxCut	-	-	5.6	4.7	25.0	**4.4**

The results of 20 replications can be seen in Table 2. We report the average absolute gap, i.e., the difference between the quality of the found solution and the best known value, of the tuned algorithms and of the state-of-the-art algorithms ITS [24], D^2TS [25], PR1 [26], and PR2 [26]. We can see that the automatic approach can find algorithms that outperform state-of-the-art algorithms. When training using Palubeckis instances, our algorithm presents an average absolute gap of 186.8, which is much better than the average of PR1. This algorithm also works well on Beasley instances, since their structure is similar to Palubeckis instances. The AAC_P algorithm is worse than the state-of-the-art on MaxCut instances, since they have a different structure. The algorithm trained with MaxCut instances presents very good results for these instances, improving slightly over PR2. However, the performance of AAC_M is worse on the other instance sets, since it is specialized on MaxCut instances. The same behavior can be observed for algorithms PR1 and PR2 of Wang et al. [26]. PR1 performs well on Beasley and Palubeckis instances, while PR2 performs well on MaxCut instances.

Algorithm 1. AAC$_R$

---

while *stopping criterion not satisfied* **do**
  $\quad E \leftarrow$ create an elite set of size e
  while *E has any novel solution* **do**
    $\quad$ **foreach** $(s,t) \in E \times E \mid s \neq t$ **do**
      $\quad\quad V \leftarrow$ variables $v \mid s[v] \neq t[v]$
      $\quad\quad d_{min} \leftarrow \gamma \times |V|$
      $\quad\quad d_{max} \leftarrow |V| - d_{min}$
      $\quad\quad d \leftarrow 0$
      $\quad\quad s^* \leftarrow s$
      $\quad\quad$ **while** $V \neq \emptyset$ **do**
        $\quad\quad\quad v \leftarrow$ select the best variable from V
        $\quad\quad\quad$ Flip $s[v]$ and remove v from V
        $\quad\quad\quad d \leftarrow d + 1$
        $\quad\quad\quad$ **if** (s *is better than* s^*) $\wedge$ ($d_{min} \leq d \leq d_{max}$) **then**
          $\quad\quad\quad\quad s^* \leftarrow s$
      $\quad\quad s \leftarrow$ iteratedTabuSearch(s^*) /* According to Algorithm 2 */
      $\quad\quad$ **if** s *is better than any solution of E* **then**
        $\quad\quad\quad$ Replace the worst solution of E by s

return *best solution of E*

4.2   Tuning with a Random Instance Set

In our second experiment, we separate the training and the validation sets. To this end, we created 20 random instances for training with a similar structure than those of the Beasley and Palubeckis instances. They have 2000 to 7000 variables and densities from 10% to 100%. The coefficients were selected uniformly at random from $[-100, 100]$.

The training process using the random instance set produced a recombination-based algorithm, which internally applies an iterated tabu search. Based on the rules presented in the grammar of Fig. 1, it consists of an algorithm based on the <RECOMBINATION> strategy, which runs the RER method with the BI strategy for the <IMPROVEMENT> step, and the ILS as the <SEARCH> step. The ILS applies a STS and the LEAST-LOSS strategy with GAMMAM step for <PERTURBATION>. We can see that the recombination strategy is used by Wang et al. [26], while the iterated tabu search is used by Palubeckis [24]. Therefore, our automatic approach combined components of both approaches. We call this algorithm AAC$_R$.

The details are shown in Algorithm 1. Repeatedly, the set of elite solutions is randomly created. Each pair of solutions is selected and then recombined using path relinking. The resulting solution is improved by an iterated tabu search, and then the elite set is updated. This process is repeated while better solutions are found and added to the elite set. The path relinking method operates on the pair of solutions s and t. It searches the space of variables that, if flipped,

---

Algorithm 2. Iterated tabu search used in AAC$_R$

---

while *stopping criterion not satisfied* **do**

$\quad$ $p_{size} \leftarrow n/g$

$\quad$ **while** $p_{size} > 0$ **do**

$\quad\quad$ $V \leftarrow$ create list with the b best variables to flip

$\quad\quad$ $v \leftarrow$ random variable from V

$\quad\quad$ Flip $s[v]$

$\quad\quad$ $p_{size} \leftarrow p_{size} - 1$

$\quad$ **while** *maximum iterations and maximum stagnation not reached* **do**

$\quad\quad$ Select variable v that leads to the best neighbor and update tabu list

$\quad\quad$ Flip $s[v]$ and set v as tabu

return *best solution found*

---

approximate s to t. Iteratively, this method selects the best variable and flips it in solution s. The new solution is accepted if it is better than the best solution found so far, and according to some minimum and maximum modification steps (d_{min} and d_{max}). The iterated tabu search procedure is detailed in Algorithm 2. It applies the least loss perturbation proposed by Palubeckis [24], iteratively flipping a random variable from the b best variables to flip. This is repeated until p_{size} variables are flipped. After perturbing the solution, the search step performs a tabu search. The best found solution is stored during the perturbation and search steps, and then returned at the final of the execution. The parameter values defined by the automatic tuning are presented in Table 3.

Table 3. Parameter values tuned by the automatic approach

Parameter	t	t_d	t_c	s	s_m	i	g	b	e	γ
Value	t_4	310	25	s_2	22	i_2	10	8	16	0.2405

The results of the execution of AAC$_R$ on the instances of Palubeckis are presented in Table 4. It presents the best and average absolute gaps obtained by AAC$_R$ and state-of-the-art approaches. We can see that we improve the results of algorithm PR1 of Wang et al. [26], presenting an average absolute gap of 211.7. AAC$_R$ also scales better and thus presents more uniform results, with the average absolute gap increasing less with the size of the instance.

Table 5 presents the results on MaxCut instances for algorithms AAC$_R$ and AAC$_M$, as well as state-of-the-art algorithms. Column "SS" shows the results for the scatter search proposed by Marti et al. [32], and column "CC" shows the results for the CirCut method proposed by Burer et al. [33]. We can see

Table 4. Best and average absolute gaps on the instances of Palubeckis.

Inst.	ITS		D^2TS		PR1		AAC_R	
	Best	Avg.	Best	Avg.	Best	Avg.	Best	Avg.
p3000-1	0.0	0.0	0.0	0.0	0.0	0.0	0.0	0.0
p3000-2	0.0	97.0	0.0	0.0	0.0	0.0	0.0	0.0
p3000-3	0.0	344.0	0.0	0.0	0.0	36.0	0.0	107.5
p3000-4	0.0	154.0	0.0	0.0	0.0	0.0	0.0	0.0
p3000-5	0.0	501.0	0.0	0.0	0.0	90.0	0.0	49.1
p4000-1	0.0	0.0	0.0	0.0	0.0	0.0	0.0	0.0
p4000-2	0.0	1285.0	0.0	0.0	0.0	71.0	0.0	323.1
p4000-3	0.0	471.0	0.0	0.0	0.0	0.0	0.0	2.7
p4000-4	0.0	438.0	0.0	0.0	0.0	0.0	0.0	0.0
p4000-5	0.0	572.0	0.0	0.0	0.0	491.0	0.0	0.0
p5000-1	700.0	971.0	325.0	656.0	0.0	612.0	0.0	386.6
p5000-2	0.0	1068.0	0.0	12533.0	0.0	620.0	0.0	338.7
p5000-3	0.0	1266.0	0.0	12876.0	0.0	995.0	0.0	76.5
p5000-4	934.0	1952.0	0.0	1962.0	0.0	1258.0	0.0	553.6
p5000-5	0.0	835.0	0.0	239.0	0.0	51.0	0.0	36.9
p6000-1	0.0	57.0	0.0	0.0	0.0	201.0	0.0	18.4
p6000-2	88.0	1709.0	0.0	1286.0	0.0	221.0	0.0	148.4
p6000-3	2729.0	3064.0	0.0	787.0	0.0	1744.0	0.0	671.9
p7000-1	340.0	1139.0	0.0	2138.0	0.0	935.0	0.0	903.4
p7000-2	1651.0	4301.0	104.0	8712.0	0.0	1942.0	8.0	828.0
p7000-3	0.0	3078.0	0.0	2551.0	0.0	332.0	0.0	0.0
Avg.	306.8	1109.6	20.4	2082.9	**0.0**	457.1	0.4	**211.7**

that AAC_M improved the results of algorithm PR2 of Wang et al. [26]. Algorithm AAC_R have worse results for MaxCut instances compared to the other approaches. This is due to the fact that the random instances have the same structure of Palubeckis and Beasley instances, but not the density and values of the MaxCut instances. Nevertheless, our results are comparable to the state-of-the-art approaches. Moreover, algorithms AAC_M and AAC_R found new best known values for the MaxCut instances, which are presented in Table 6.

Table 5. Average absolute gap on the MaxCut instances.

Inst.	PR2	SS	CC	AAC$_M$	AAC$_R$	Inst.	PR2	SS	CC	AAC$_M$	AAC$_R$
G1	0.0	0.0	0.0	0.0	16.2	G28	7.1	11.0	36.0	8.3	10.8
G2	0.0	0.0	3.0	0.0	13.8	G29	13.1	16.0	29.0	14.8	21.6
G3	2.0	0.0	0.0	0.0	10.8	G30	7.2	9.0	27.0	6.5	14.2
G4	0.0	0.0	5.0	0.0	12.8	G31	6.5	18.0	21.0	5.3	12.2
G5	0.0	0.0	4.0	0.0	10.9	G32	5.4	12.0	20.0	8.7	34.6
G6	0.0	13.0	0.0	0.0	5.9	G33	5.9	20.0	22.0	8.7	29.8
G7	0.0	24.0	3.0	0.0	10.2	G34	5.8	20.0	16.0	8.0	31.7
G8	0.0	19.0	2.0	0.0	6.5	G35	13.2	16.0	14.0	15.2	41.5
G9	0.0	140	6.0	0.0	9.6	G36	18.3	17.0	17.0	17.0	43.5
G10	0.2	7.0	6.0	0.0	10.2	G37	21.1	25.0	23.0	19.0	44.0
G11	0.0	2.0	4.0	0.0	7.6	G38	11.6	1.0	36.0	10.3	39.5
G12	0.0	4.0	4.0	0.0	10.0	G39	15.9	14.0	12.0	11.8	65.5
G13	0.0	4.0	8.0	0.0	14.0	G40	15.7	25.0	12.0	11.2	74.1
G14	1.4	4.0	6.0	1.4	8.1	G41	15.1	18.0	6.0	11.0	74.8
G15	0.7	1.0	1.0	0.1	14.2	G42	11.8	21.0	9.0	10.2	73.1
G16	0.6	7.0	7.0	0.1	13.6	G43	0.1	4.0	4.0	0.0	10.7
G17	0.6	4.0	10.0	0.3	12.7	G44	0.1	2.0	7.0	0.0	8.6
G18	0.0	4.0	14.0	0.0	17.7	G45	0.1	12.0	2.0	0.0	6.6
G19	0.0	3.0	18.0	0.0	19.1	G46	0.2	15.0	4.0	0.0	7.5
G20	0.0	0.0	0.0	0.0	25.9	G47	0.2	8.0	1.0	0.1	9.0
G21	0.0	1.0	0.0	0.0	27.5	G48	0.0	0.0	0.0	0.0	0.0
G22	4.5	13.0	13.0	9.8	20.6	G49	0.0	0.0	0.0	0.0	0.0
G23	10.4	25.0	25.0	9.9	12.0	G50	0.0	0.0	0.0	0.0	5.0
G24	11.7	34.0	23.0	15.1	15.9	G51	1.6	2.0	11.0	1.4	15.6
G25	10.8	19.0	13.0	9.8	14.1	G52	2.6	2.0	18.0	1.1	14.9
G26	13.7	32.0	12.0	11.2	14.3	G53	2.3	4.0	8.0	2.0	14.8
G27	10.1	19.0	31.0	8.8	20.8	G54	4.2	6.0	10.0	1.9	15.7
						Avg.	4.7	10.2	10.8	**4.4**	20.3

Table 6. New best known values found.

Instance	G25	G26	G27	G28	G30	G31	G38
Previous value	13339	13326	3337	3296	3412	3306	7682
New value	13340	13327	3341	3298	3413	3310	7383

5   Conclusions

This paper presents the application of a grammar-based approach to the automatic design of heuristic algorithms for the unconstrained binary quadratic programming. The grammar describes heuristic components, including those used by state-of-the-art approaches, and the related parameters. We then use iterated F-race to explore the search space defined by the grammar and find good algorithms. Our results show the potential of automatic algorithm configuration for designing heuristic methods. It is an effective tool for the trial-and-error process of searching for a good algorithm, reducing human effort in those tasks. This potential can be seen in the AAC_R algorithm, which combines the components from different algorithms of literature into a new, competitive algorithm.

The specialized algorithms AAC_P and AAC_M present very good results, outperforming state-of-the-art approaches in the Palubeckis and MaxCut instance sets, respectively. The algorithm trained with the random instances also performs well on Palubeckis instances, and is comparable on MaxCut instances. We also note that our algorithms are not dominated by any of the algorithms from the literature. As future work, we aim at find a single algorithm that is better than the state-of-the-art approaches on both instance sets.

References

1. Hutter, F., Hoos, H.H., Leyton-Brown, K., Stützle, T.: ParamILS: an automatic algorithm configuration framework. J. Artif. Intell. Res. **36**(1), 267–306 (2009)
2. Ansótegui, C., Sellmann, M., Tierney, K.: A gender-based genetic algorithm for the automatic configuration of algorithms. In: Gent, I.P. (ed.) CP 2009. LNCS, vol. 5732, pp. 142–157. Springer, Heidelberg (2009). https://doi.org/10.1007/978-3-642-04244-7_14
3. Hutter, F., Hoos, H.H., Leyton-Brown, K.: Sequential model-based optimization for general algorithm configuration. In: Coello, C.A.C. (ed.) LION 2011. LNCS, vol. 6683, pp. 507–523. Springer, Heidelberg (2011). https://doi.org/10.1007/978-3-642-25566-3_40
4. López-Ibáñez, M., Dubois-Lacoste, J., Cáceres, L.P., Birattari, M., Stützle, T.: The irace package: iterated racing for automatic algorithm configuration. Oper. Res. Perspect. **3**, 43–58 (2016)
5. Alidaae, B., Kochenberger, G.A., Ahmadian, A.: 0–1 quadratic programming approach for optimum solutions of two scheduling problems. Int. J. Syst. Sci. **25**(2), 401–408 (1994)
6. Hansen, P., Jaumard, B.: Algorithms for the maximum satisfiability problem. Computing **44**(4), 279–303 (1990)
7. Pardalos, P.M., Xue, J.: The maximum clique problem. J. Global Optim. **4**(3), 301–328 (1994)
8. Pardalos, P.M., Jha, S.: Complexity of uniqueness and local search in quadratic 0–1 programming. Oper. Res. Lett. **11**(2), 119–123 (1992)
9. Kochenberger, G.A., Glover, F., Alidaee, B., Rego, C.: A unified modeling and solution framework for combinatorial optimization problems. OR Spectr. **26**, 237–250 (2004)

10. Ansótegui, C., Malitsky, Y., Samulowitz, H., Sellmann, M., Tierney, K.: Model-based genetic algorithms for algorithm configuration. In: IJCAI, pp. 733–739 (2015)
11. Hutter, F., Hoos, H.H., Leyton-Brown, K.: Parallel algorithm configuration. In: Hamadi, Y., Schoenauer, M. (eds.) LION 2012. LNCS, pp. 55–70. Springer, Heidelberg (2012). https://doi.org/10.1007/978-3-642-34413-8_5
12. Birattari, M., Stützle, T., Paquete, L., Varrentrapp, K.: A racing algorithm for configuring metaheuristics. In: 4th Annual Conference on Genetic and Evolutionary Computation, pp. 11–18. Morgan Kaufmann Publishers Inc. (2002)
13. KhudaBukhsh, A.R., Xu, L., Hoos, H.H., Leyton-Brown, K.: SATenstein: automatically building local search SAT solvers from components. Artif. Intell. **232**, 20–42 (2016)
14. López-Ibáñez, M., Stützle, T.: The automatic design of multiobjective ant colony optimization algorithms. IEEE Trans. Evol. Comput. **16**(6), 861–875 (2012)
15. Bezerra, L.C.T., López-Ibáñez, M., Stützle, T.: Automatic design of evolutionary algorithms for multi-objective combinatorial optimization. In: Bartz-Beielstein, T., Branke, J., Filipič, B., Smith, J. (eds.) PPSN 2014. LNCS, vol. 8672, pp. 508–517. Springer, Cham (2014). https://doi.org/10.1007/978-3-319-10762-2_50
16. O'Neill, M., Ryan, C.: Grammatical evolution. IEEE Trans. Evol. Comput. **5**(4), 349–358 (2001)
17. Burke, E.K., Hyde, M.R., Kendall, G.: Grammatical evolution of local search heuristics. IEEE Trans. Evol. Comput. **16**(3), 406–417 (2012)
18. Tavares, J., Pereira, F.B.: Automatic design of ant algorithms with grammatical evolution. In: Moraglio, A., Silva, S., Krawiec, K., Machado, P., Cotta, C. (eds.) EuroGP 2012. LNCS, vol. 7244, pp. 206–217. Springer, Heidelberg (2012). https://doi.org/10.1007/978-3-642-29139-5_18
19. Rothlauf, F., Oetzel, M.: On the locality of grammatical evolution. In: Collet, P., Tomassini, M., Ebner, M., Gustafson, S., Ekárt, A. (eds.) EuroGP 2006. LNCS, vol. 3905, pp. 320–330. Springer, Heidelberg (2006). https://doi.org/10.1007/11729976_29
20. Lourenço, N., Pereira, F.B., Costa, E.: Unveiling the properties of structured grammatical evolution. Genet. Program. Evol. Mach. **17**(3), 251–289 (2016)
21. Mascia, F., López-Ibáñez, M., Dubois-Lacoste, J., Stützle, T.: From grammars to parameters: automatic iterated greedy design for the permutation flow-shop problem with weighted tardiness. In: Nicosia, G., Pardalos, P. (eds.) LION 2013. LNCS, vol. 7997, pp. 321–334. Springer, Heidelberg (2013). https://doi.org/10.1007/978-3-642-44973-4_36
22. Mascia, F., López-Ibáñez, M., Dubois-Lacoste, J., Stützle, T.: Grammar-based generation of stochastic local search heuristics through automatic algorithm configuration tools. Comput. Oper. Res. **51**, 190–199 (2014)
23. Kochenberger, G., Hao, J.K., Glover, F., Lewis, M., Lü, Z., Wang, H., Wang, Y.: The unconstrained binary quadratic programming problem: a survey. J. Comb. Optim. **28**(1), 58–81 (2014)
24. Palubeckis, G.: Iterated tabu search for the unconstrained binary quadratic optimization problem. Informatica **17**(2), 279–296 (2006)
25. Glover, F., Lü, Z., Hao, J.K.: Diversification-driven tabu search for unconstrained binary quadratic problems. 4OR: Q. J. Oper. Res. **8**(3), 239–253 (2010)
26. Wang, Y., Lü, Z., Glover, F., Hao, J.K.: Path relinking for unconstrained binary quadratic programming. EJOR **223**(3), 595–604 (2012)
27. Glover, F.: Tabu search. ORSA J. Comput. **1**(3), 190–206 (1989)

28. Lourenço, H.R., Martin, O.C., Stützle, T.: Iterated local search. In: Gendreau, M., Potvin, J.Y. (eds.) Handbook of Metaheuristics. International Series in Operations Research and Management Science, vol. 146, pp. 321–354. Springer, Heidelberg (2003). https://doi.org/10.1007/978-1-4419-1665-5_12

29. Merz, P., Freisleben, B.: Greedy and local search heuristics for unconstrained binary quadratic programming. J. Heuristics **8**(2), 197–213 (2002)

30. Feo, T.A., Resende, M.G.: Greedy randomized adaptive search procedures. J. Global Optim. **6**(2), 109–133 (1995)

31. Hyde, M.R., Burke, E.K., Kendall, G.: Automated code generation by local search. J. Oper. Res. Soc. **64**(12), 1725–1741 (2013)

32. Martí, R., Duarte, A., Laguna, M.: Advanced scatter search for the max-cut problem. INFORMS J. Comput. **21**(1), 26–38 (2009)

33. Burer, S., Monteiro, R.D., Zhang, Y.: Rank-two relaxation heuristics for max-cut and other binary quadratic programs. SIAM J. Optim. **12**(2), 503–521 (2002)

Automatic Algorithm Configuration for the Permutation Flow Shop Scheduling Problem Minimizing Total Completion Time

Artur Brum[(⊠)] and Marcus Ritt

Instituto de Informática, Universidade Federal do Rio Grande do Sul (UFRGS),
Porto Alegre, Brazil
{artur.brum,marcus.ritt}@inf.ufrgs.br

Abstract. Automatic algorithm configuration aims to automate the often time-consuming task of designing and evaluating search methods. We address the permutation flow shop scheduling problem minimizing total completion time with a context-free grammar that defines how algorithmic components can be combined to form a full heuristic search method. We implement components from various works from the literature, including several local search procedures. The search space defined by the grammar is explored with a racing-based strategy and the algorithms obtained are compared to the state of the art.

Keywords: Automatic algorithm configuration
Iterated greedy algorithm · Iterated local search
Flow shop scheduling problem · Total completion time

1 Introduction

The design of heuristic search methods is a complex problem that requires significant time and effort and, therefore, can benefit from an automated approach that is both faster and more systematic. Automatic algorithm configuration (AAC) comprises the automated selection and calibration of algorithmic components for a given problem and aims to automatize such a task, allowing the researchers to focus more on component design decisions and turning the whole process more robust and less error-prone.

AAC has received an increasing attention recently and is applied to a wide assortment of problems, e.g. the bi-objective traveling salesman problem [1], the one-dimensional bin packing problem [2] and the propositional satisfiability problem [3]. In [4] a literature review on AAC for production scheduling problems is presented.

A. Brum—CNPq–Brazil scholarship holder.

A. Liefooghe and M. López-Ibáñez (Eds.): EvoCOP 2018, LNCS 10782, pp. 85–100, 2018.
https://doi.org/10.1007/978-3-319-77449-7_6

In this work we address the permutation flow shop scheduling problem (PFSSP) with total completion time minimization. The PFSSP is a well-known optimization problem that consists in scheduling a set of n jobs that have to be processed in the same order on m machines. Each job must be processed on all machines and each machine can process at most one job at a time. Since all machines process the jobs in the same order, solutions for the PFSSP can be represented by permutations of jobs. We focus on the minimization of the total completion time, one of the most common objective functions in the literature. Defining C_{ij} as the completion time of job j on machine i, the total completion time is denoted by $C_{sum} = \sum_{j=1}^{n} C_{mj}$. The PFSSP with total completion time minimization is denoted as $F|prmu|\sum C_j$ according to [5].

As the PFSSP for total completion time is NP-hard for $m \geq 2$ [6], heuristic methods are frequently adopted and many different approaches can be found in the literature. Recently, the PFSSP has been addressed by means of AAC in [7], which uses genetic programming (GP) to improve the Nawaz et al. [8] constructive heuristic. The replacement of the evolutionary algorithm in grammatical evolution (GE) with a parameter configuration tool is proposed in [9]. The strategy is evaluated in the configuration of an iterated greedy algorithm for the PFSSP with weighted tardiness minimization. In a follow-up work, [10] presents an iterated-local-search-based generalized structure that incorporates several classic local search methods for the same problem.

We follow a strategy similar to the one used in [9], implementing several algorithmic components from the literature and proposing a grammar that determines how to combine them. We adopt a solution representation based on the structured grammatical approach proposed by [11], which is also similar to the approach of [9], but does not use conditional parameters. Our objective in this work is to compare algorithms constructed via AAC to the state-of-the-art methods regarding total completion time minimization.

Our methodology is presented and detailed in Sect. 2. The computational experiments and results are described in Sect. 3, and we present our conclusions in Sect. 4.

2   Automatic Algorithm Configuration

A significant part of the published research regarding AAC employs methods such as ParamILS [12], SMAC [13], irace [14] or evolutionary approaches, such as GP or GE. In GE, a context-free grammar is used in order to specify how algorithmic components like local search procedures, perturbation functions and acceptance criteria can be combined to form a full heuristic search method. To explore the search space defined by the grammar, a genetic algorithm is employed in which each individual corresponds to an instantiation of such a grammar, i.e. a heuristic search method. The genotype of each individual is represented by an array of integer values called codons. The codon value is usually used with the modulo operator to select a production rule for a given non-terminal.

However, this strategy has some issues related to its high redundancy and low locality [15], i.e. one specific algorithm can be represented by several different genotypes, and even a slight change in one codon of a certain genotype can induce a vastly different phenotype. Structured grammatical evolution (SGE) is an approach proposed by [11] in order to address those issues by defining a hierarchical structure and linking non-terminals to specific elements. The solution representation in SGE is adopted in this work, along with the search strategy of [9], which incorporates a mapping to a parametric representation and the use of an automatic parameter configuration tool as an alternative to the evolutionary algorithm in SGE. We start by presenting our proposed grammar, describing its components and then detailing our solution representation and search strategy.

2.1   Grammar and Components

We adopt the grammar presented in Backus-Naur Form (BNF) in Fig. 1. The components represent different heuristic search strategies, constructive heuristics to generate initial permutations, tie-breakers, local search procedures and perturbation functions. Numeric parameters are not shown for the sake of simplicity. This grammar is biased towards algorithms that are known to be effective for the PFSSP, different from approaches such as [10], which aim to represent a broader range of heuristic strategies, in order to reduce the search effort.

```
 1  <START>      ::= iga(<INI_SOL>,<TIEBREAK>,<PARTIAL>,<LS>) |
 2                   ils(<INI_SOL>,<LS>,<PERTURB>)
 3  <INI_SOL>    ::= nehCsum | LR | BSCH | FRB5(<TIEBREAK>)
 4  <TIEBREAK>   ::= KK1 | KK2 | First | Last | Random
 5  <PARTIAL>    ::= ε | insertion(<TIEBREAK>)
 6  <LS>         ::= <LS_PROC> | alternate(<LS_PROC>,<LS_PROC>)
 7  <LS_PROC>    ::= insertion(<TIEBREAK>) | swapTasgetiren |
 8                   swapInc | insertTasgetiren |
 9                   lsTasgetiren | fpe | bpe | iRZ |
10                   riRZ | raiRZ | insertFPR | swapFirst |
11                   swapBest | swapR | insertJarboui
12  <PERTURB>    ::= swap | shift
```

Fig. 1. Grammar which defines how the components can be combined.

Search Strategy. For the top-level search mechanism, we selected two state-of-the-art metaheuristics from the literature: the iterated greedy algorithm (IGA) for the PFSSP proposed in [16] and the Multi-Restart Iterated Local Search (MRSILS) of [17].

The IGA for the PFSSP is a simple algorithm composed by two main phases that can be summarized as follows. The first phase, named destruction, removes d

---

Algorithm 1. Insertion local search for the PFSSP

Input: Permutation π
Output: Best permutation found π^*
1: **function** INSERTION(π)
2:    **repeat**
3:       improved = false
4:       $\pi' = \pi$
5:       **for** $k = 1$ to n **do**
6:          $j = k^{th}$ job in π'
7:          $\pi' = $ REMOVE(j,π')
8:          $\pi' = $ INSERTBESTPOSITION(j,π')
9:          **if** $f(\pi') \leq f(\pi)$ **then**
10:             $\pi = \pi'$
11:             improved = true
12:          **end if**
13:       **end for**
14:    **until** not improved
15:    **return** π^*
16: **end function**

---

random jobs from the current permutation. The removed jobs are then reinserted greedily into the schedule during the second phase, named construction. This process is then repeated until a certain termination criterion is met.

Additionally, a local search procedure is added after construction in order to improve the permutation obtained in that phase. The proposed method implements an insertion neighborhood: each job is removed and reinserted at the position that leads to the smallest objective function value. The procedure is similar to `insertion` on lines 5 and 7 of the grammar in Fig. 1 and is presented in Algorithm 1, where $f(\pi)$ denotes the objective function value of permutation π. All presented algorithms implicitly maintain the best permutation visited so far π^*.

As observed in [16], several alternative local search procedures can be found in the literature. In this work we consider 15 distinct methods.

After the local search, a permutation is accepted according to a Metropolis criterion [18], i.e. it is always accepted if it improves the objective, otherwise it is accepted with a probability that decreases as the objective difference increases, as shown in Algorithm 2. The temperature value is defined in [16] as:

$$T = \alpha \times \frac{\sum_{i=1}^{m} \sum_{j=1}^{n} p_{ij}}{10nm} \tag{1}$$

Here p_{ij} is the processing time of job j on machine i and α is a parameter.

The IGA is outlined in Algorithm 3. Besides providing state-of-the-art permutations, it is worth to highlight other positive characteristics such as having only two parameters (namely d and α), simplicity, making it easy to implement and favoring the reproduction of published results, and adaptability, as the method can be applied to other flow shop variants [16].

Algorithm 2. Acceptance criterion

Input: Current permutation π, new permutation π', temperature T
Output: Permutation π
1: **function** ACCEPT(π, π', T)
2:     $r = $ RANDOM$(0, 1)$ ▷ random number in $[0,1)$
3:     $\Delta = f(\pi') - f(\pi)$
4:     **if** $\Delta \leq 0$ or $r \leq e^{-\Delta/T}$ **then**
5:         $\pi = \pi'$
6:     **end if**
7:     **return** π
8: **end function**

Algorithm 3. IGA for the PFSSP

Input: Integer d, temperature T
Output: Best permutation found π^*
1: **function** IGA(d,T)
2:     $\pi = $ INITIALPERMUTATION
3:     **while** termination criterion not met **do**
4:         $\pi' = $ DESTRUCT(π, d)
5:         $\pi' = $ CONSTRUCT(π')
6:         $\pi' = $ LOCALSEARCH(π')
7:         $\pi = $ ACCEPT(π, π', T)
8:     **end while**
9:     **return** π^*
10: **end function**

In another work related to the IGA for the PFSSP, [19] proposes the addition of a local search to improve partial permutations. The same procedure described in Algorithm 1 is applied to optimize the partial permutation obtained after the destruction phase. The results showed that such a strategy is able to improve the makespan and contributed to the improvement of the state of the art. We therefore evaluate it for total completion time minimization in this work. The strategy is represented by the <PARTIAL> non-terminal in the grammar. It is optional and specific to the IGA.

The other metaheuristic framework adopted in this work is similar to MRSILS, proposed in [17]. Iterated local search (ILS) is a metaheuristic that repeatedly applies a local search and slightly perturbates the current solution between iterations. This perturbation is usually probabilistic and aims to assist the search escaping local optima. The MRSILS includes a pool of permutations, which is used to restart the search once it is unable to improve the incumbent permutation for a certain number of iterations. Algorithm 4 presents the method. It was adapted to support the multiple local search procedures we consider in this work, as it was presented with an insertion local search. The functions in lines 6, 9 and 11–14 are straightforward: Perturb(π) applies a perturbation to permutation π, ClearPool empties the pool of permutations, NotInPool(π') returns true if π' is not in the pool and false otherwise, InsertIntoPool(π') adds permutation

---

Algorithm 4. ILS with a pool of permutations for the PFSSP

Input: Integer p, integer ps
Output: Best permutation found π^*
 1: **function** ILS(p,ps)
 2:　　$\pi = $ INITIALPERMUTATION
 3:　　$\pi = $ LOCALSEARCH(π)
 4:　　$\pi^* = \pi$
 5:　　**while** termination criterion not met **do**
 6:　　　　$\pi' = $ PERTURB(π, p)
 7:　　　　$\pi' = $ LOCALSEARCH(π')
 8:　　　　**if** UPDATEDINCUMBENT **then**
 9:　　　　　　CLEARPOOL
10:　　　　**end if**
11:　　　　**if** NOTINPOOL(π') **then**
12:　　　　　　INSERTINTOPOOL(π')
13:　　　　　　**if** POOLSIZE $> ps$ **then**
14:　　　　　　　　REMOVEWORSTFROMPOOL
15:　　　　　　**end if**
16:　　　　**end if**
17:　　　　$\pi = $ SELECTFROMPOOL
18:　　**end while**
19:　　**return** π^*
20: **end function**

π' to the pool, PoolSize returns the number of permutations currently in the pool and RemoveWorstFromPool removes the permutation with worst objective function value from the pool. Function SelectFromPool in line 16 determines the current permutation according to the following strategy: if the permutation pool is not full then return the best one in it, otherwise return a randomly selected permutation from the pool. With that established, note that the pool size is equivalent to the number of iterations without improvement after which restarts from random permutations from the pool are done.

The perturbation function performs p random movements and is specific to the ILS. It is represented by the <PERTURB> non-terminal in the grammar and has two derivation options that determine if the movements performed are shift or swap movements. A shift movement reinserts a given job in a different position, while a swap movement exchanges the positions of two jobs.

Initial Permutation. We implemented four constructive heuristics to generate an initial permutation: NEH for total completion time [20], LR [21], the strategy FRB5 [22] adapted to minimize completion time, and BSCH [23].

The NEH procedure for total completion time minimization can be summarized as follows. The n jobs are sorted in a non-decreasing order according to the sum of their processing times across all m machines. Now, according to this order, each job is inserted in the position that yields the minimum total completion time. This step is repeated until all jobs are inserted. FRB5 works similarly, with the addition of an insertion local search (see Algorithm 1) that is performed after each job is added to the schedule.

The LR constructive heuristic starts by ordering the jobs according to an indicator that considers both the weighted idle time induced when appending a job j at the end of the schedule, and an estimation for the completion time of jobs appended after j. Then, x schedules are created each starting with one of the first x jobs of this order, and the remaining jobs are appended one at a time using the same indicator. The final permutation is the best one out of the x generated permutations. We fix $x = \lceil n/m \rceil$, which is the best value found in [21].

Finally, BSCH is a more recent beam search heuristic. Similar to LR, the method appends jobs at the end of w partial sequences. The underlying idea is similar to a branch-and-bound algorithm, but only a reduced number of nodes is kept at each level. This number is defined by a parameter commonly called the beam width (w). In each level and for each node, a job is appended at the end of the partial sequence. To select such jobs, a forecast index which considers not only the current partial schedule but also the unassigned jobs is used. After evaluating all the candidates, only the w best nodes are kept and this is repeated until the nodes are complete permutations, in which case the best one is selected. We set $w = n$, as it performed best according to [23].

Local Search. It was observed in [24] that some local search procedures often complement each other, therefore it was adopted a strategy in which two different procedures are applied alternately. In their case, a swap local search is performed if the current iteration is even, and a shift local search is performed otherwise. We implement an analogous strategy in this work. The `<LS>` non-terminal in Fig. 1 can be derived into a single local search or an alternation between two local searches. In this section we present a brief description for each of the 15 distinct local search methods we consider.

- `insertion`: remove each job and reinsert it in the best position. Repeat this process until no improvement is found (see Algorithm 1). Allows the use of tie-breaking functions.
- `insertJarboui`: insertion local search of [25]. Reinsert each job in all positions, immediately accepting improving movements and repeating the search until a local minimum is obtained. Slightly different from `insertion` regarding the selection of jobs to reinsert.
- `swapTasgetiren`: swap local search of [26]. Swaps all pairs of jobs, restarting the search when an improvement is found.
- `insertTasgetiren`: insertion local search of [26]. Reinsert every job j in all positions after the current position of j. The search is restarted when an improvement is found.
- `lsTasgetiren`: another local search method of [26]. Repeatedly apply insert-Tasgetiren followed by swapTasgetiren until a local minimum is obtained.

- insertFPR: insertion local search of [26], very similar to insertion. In a cyclic order, each job is removed and reinserted in the best position. This is repeated until there is no improvement for n consecutive insertions.
- fpe: forward pairwise exchange local search of [21]. From an initial permutation, exchange each job j with the x jobs scheduled after j. A movement is accepted if it improved the permutation, otherwise it is immediately undone. The value of x is set to $\lceil n/m \rceil$.
- bpe: backward pairwise exchange local search of [21]. Similar to fpe, except for the exchange order, which is reversed, i.e. each job is exchanged with the x jobs scheduled prior to it. The value of x is defined as in fpe.
- iRZ: insertion local search of [27], which evaluates the insertion of all jobs in all positions. Similar to insertion and insertJarboui, but with a slightly different behavior when improvements are found.
- riRZ: same as iRZ, except for the order jobs are evaluated, which is reversed, i.e. starts with the last job of the schedule.
- raiRZ: same as iRZ, except for the order jobs are evaluated, which is random.
- swapInc: swap local search of [24], which exchanges the jobs in positions i and $i + q$, immediately accepting improving swaps. If none of the swaps improves the current permutation, q is incremented and the search is restarted. Otherwise, when an improvement is found, q is set to a value q_{min}. The search stops when $q > q_{max}$ or a total number of swaps s_{max} is performed. According to [24], we set $q_{min} = 1$, $q_{max} = n$ and $s_{max} = 3n^2$.
- swapFirst: cyclically swap every pair of adjacent jobs with a first-improvement strategy. The search stops when $n_{ls} \cdot (n-1)$ swaps are evaluated, where n_{ls} is a parameter.
- swapBest: similar to swapFirst, with a best-improvement strategy.
- swapR: for each pair of jobs (i, j), insert j at the position of i and reinsert i optimally. Inspired by [28].

Regarding the n_{ls} parameter, it not only limits the number of swaps in swapFirst and swapBest, but also defines a maximum number of times the full neighborhood is scanned in insertion, iRZ, riRZ and raiRZ. This is adopted as [29] observed that a local minimum is often found with only a few scans in similar procedures, thus establishing such a limit can help to improve the performance. We do not implement this limit for the other procedures since they were not proposed with such a mechanism in the literature.

Tie-Breakers. Finally, when evaluating the possible insertion positions for a certain job, we often observe multiple positions that lead to the same objective function value. In these cases, we break ties by one of five different rules:

- First: select the position of smallest index.
- Last: select the position of biggest index.
- KK1: tie-breaking rule of [30].
- KK2: tie-breaking rule of [31].
- Random: select a random position [32].

2.2 Solution Representation

In SGE an individual is represented by a sequence of genes, each one containing a list of integer values. Given the grammar in Fig. 1 we have 7 non-terminals (<START>, <INI_SOL>, <TIEBREAK>, <PARTIAL>, <LS>, <LS_PROC> and <PERTURB>), so the SGE genotype contains 7 genes. Now, to determine the length of the list of each gene we observe the maximum number of expansions of the respective non-terminal. To determine the interval each value in the list can assume, we count the number of derivation options of each non-terminal, thus we have:

- <START>: is expanded only once, so the list is of size 1. The value is in the interval $[0, 1]$, as there are 2 derivation options (iga or ils).
- <INI_SOL>: also expanded only once and has 4 derivation options, resulting in a list of size 1 and interval $[0, 3]$.
- <TIEBREAK>: can be expanded up to 5 times: one if iga is selected in line 1, one if the initial permutation is generated with FRB5, another if the partial permutation is optimized and two times if two local search procedures are selected and derived to insertion. With 5 derivation options, the interval is $[0, 4]$.
- <PARTIAL>: expanded once with two derivation options, thus list of size 1 and interval $[0, 1]$.
- <LS>: is expanded only once and there are only 2 derivation options, resulting in a list of size 1 and an interval equals to $[0, 1]$.
- <LS_PROC>: can be expanded up to 2 times and has 15 derivation options, so the list is of size 2 and the interval is $[0, 14]$.
- <PERTURB>: another non-terminal that is expanded at most once and has two derivation options, resulting in a list of size 1 and interval $[0, 1]$.

An example of genotype is $g = [\{0\}, \{3\}, \{0, 2, 4, 4, 0\}, \{1\}, \{1\}, \{0, 13\}, \{0\}]$. To translate it into an algorithm, we start from the left-most gene ($\{0\}$). It is associated to the first non-terminal in the grammar, which is <START>. The first value in the gene corresponds to the first time the non-terminal is expanded. As mentioned earlier, this gene has only one value, as <START> is expanded only one time. Now, we assign the left-most derivation option, in this case iga, to the left-most value of the interval, which is 0. The second option, ils, is assigned to the second value, 1. Therefore, since the first gene has value 0, g translates into an IGA.

The second gene corresponds to the second non-terminal, INI_SOL, and, like the previous one, only has one value. We see that the value 3 translates into FRB5. The third gene is related to tie-breaking rules. The <TIEBREAK> non-terminal can be expanded up to 5 times, in the following order, when:

1. IGA is selected.
2. Initial permutation is generated by FRB5.
3. Local search is applied to the partial permutation.
4. First local search procedure is insertion.
5. Second local search procedure is insertion.

Thus, the tie-breaking rules are, respectively: KK1, first, random, random and KK1.

Both the fourth and fifth gene contain a single value and decide, respectively, that the local search is to be applied to the partial permutation (value is 1) and that two local search procedures should be adopted (1 indicates two algorithms). The procedures are defined by the values in the sixth gene, which select insertion (with random tie-breaker) and swapR.

Finally, the seventh and last gene corresponds to a perturbation function that performs swaps, however, it is irrelevant in this case since it is specific to ILS. We therefore conclude that g corresponds to an IGA with partial permutation optimization, with an initial permutation built by FRB5 and that alternates between insertion and swapR during the local search phase.

2.3   Search Strategy

Having established a grammar that determines how individual components can be combined to generate an algorithm and having defined an approach to represent instantiations of the grammar, we now focus on a search method to explore the search space defined by the grammar. We map a genotype to a sequence of categorical parameters and employ the state-of-the-art parameter configuration tool irace in order to find good configurations. irace implements Elitist Iterated F-Race [33], a racing procedure that applies the Friedman test to discard inferior configurations.

Besides the genotype, other parameters associated to the phenotype must be configured as well. If the algorithm is an IGA, then the number of jobs removed in the destruction phase d and the temperature multiplier α have to be calibrated. For the ILS we have the pool size ps and the number of moves performed by the perturbation procedure p. We used ranges including typical values used in the literature: $d \in [1, 20]$, $\alpha \in [0.01, 1]$, $p \in [1, 20]$ and $ps \in [2, 20]$. Finally, we establish that $n_{ls} \in [1, 20]$ and define it conditional to at least one of the following iterative improvement local search procedures being selected: insertion, iRZ, riRZ, raiRZ, swapFirst and swapBest.

3   Computational Experiments

3.1   Benchmarks

In this work we consider the well-known benchmarks of [34,35]. The former contains 12 groups of 10 instances of the same size, while the latter contains 48 groups, also with 10 instances of the same size.

3.2   Experimental Setup

We execute irace 10 times, each with its own set of randomly generated training instances in order to avoid overtuning. Each set contains 120 instances with the same dimensions as in the Taillard benchmark and with uniform random processing times in the interval $[1, 99]$. Furthermore, each `irace` run is restricted to a budget of 2400 candidate runs and each candidate run is limited to $10\,nm$ milliseconds. The best algorithm found in each run is then evaluated on the Taillard benchmark.

The experiments were executed on a PC with an Intel Core i7 930 processor and 12 GB of main memory, running Ubuntu 16.04.3. Our method was implemented in C++ and compiled with `g++` 5.4.0 using `-O3` flag. In `irace` (version 2.4.1844) we allowed the parallel evaluation of candidates, limited to 4 threads.

3.3   Results

First, we present the best algorithm of each `irace` run in Table 1. The columns present the algorithm name ("Alg."), the chosen metaheuristic ("M"), the constructive heuristic to generate an initial permutation ("C.H."), the tie-breaker for the reinsertion phase in IGA ("IGA"), whether local search for partial solutions is applied or not ("Partial"), the first and the second local search procedures ("Proc. 1", "Proc. 2"), and the numeric parameters d, n_{ls} and α.

Table 1. Algorithms found by the 10 runs of irace.

Alg.	M	C.H	Tie-breaker IGA	Local search Partial	Proc. 1	Proc. 2	d	n_{ls}	α
A_0	IGA	BSCH	KK1	No	riRZ	-	6	9	0.2
A_1	IGA	BSCH	KK1	No	insertFPR	-	11	-	0.1
A_2	IGA	BSCH	First	No	riRZ	fpe	8	19	0.1
A_3	IGA	BSCH	KK2	No	raiRZ	-	9	15	0.1
A_4	IGA	BSCH	KK2	No	iRZ	-	8	15	0.1
A_5	IGA	BSCH	Random	No	fpe	-	7	-	0.2
A_6	IGA	BSCH	KK2	No	swapTasgetiren	riRZ	5	16	0.5
A_7	IGA	BSCH	KK1	No	swapInc	insertFPR	7	-	0.3
A_8	IGA	BSCH	First	No	fpe	-	6	-	0.1
A_9	IGA	BSCH	Last	No	bpe	iRZ	7	7	0.1

Analyzing the obtained algorithms we observe that all of them are iterated greedy algorithms and all did select BSCH as the constructive heuristic, which is to be expected since the results reported in [23] show a significant advantage over other constructive heuristics. Another choice common to all algorithms is the exclusion of the local search for partial permutations, which suggests that this strategy is less useful for total completion time minimization and may be more effective for minimizing makespan, for which it was proposed. The most frequently selected local search procedures are iRZ and its variants (riRZ and `raiRZ`) and `insertFPR`. The choice of tie-breakers shows a significant variation: KK1 and KK2 are selected more frequently, but all rules are selected at least once. Regarding the numeric parameters, the values for d in $[5, 11]$ confirm previous findings that removing and reinserting a higher number of jobs is beneficial when minimizing completion time [24]. The high values for n_{ls} are mostly equivalent, since the local search usually terminates earlier in a local minimum. The α-values are very similar, except when the local searches include non-adjacent swaps.

We evaluate algorithms A_0 to A_9 on the Taillard benchmark and compare the results to the state of the art metaheuristic MRSILS(BSCH) of [23]. The results are presented in Table 2 as the average relative percentage deviation (ARPD) from the upper bounds reported by [36]. For a fair comparison we reimplemented MRSILS(BSCH) and present the obtained results in column "MRSILS". The parameter values used for MRSILS(BSCH) are the same as in [17]. For each algorithm 10 replications per instance were performed, each one with a time limit of $30\,nm$ milliseconds. Columns "A_0" to "A_9" show the individual results, and column "Avg." the average result of the 10 algorithms. Note that some values are negative since the results improve the upper bounds.

Table 2. ARPD for MRSILS(BSCH) and the 10 algorithms obtained via AAC.

Inst.	MRSILS	A_0	A_1	A_2	A_3	A_4	A_5	A_6	A_7	A_8	A_9	Avg.
20×5	0.007	0.002	0.001	0.004	0.006	0.006	0.000	0.000	0.000	0.006	0.005	0.003
20×10	0.000	0.000	0.000	0.000	0.000	0.000	0.001	0.000	0.000	0.005	0.000	0.001
20×20	0.000	0.000	0.000	0.000	0.000	0.000	0.002	0.000	0.000	0.010	0.000	0.001
50×5	0.173	0.219	0.178	0.152	0.173	0.162	0.177	0.248	0.190	0.141	0.153	0.179
50×10	0.551	0.464	0.536	0.564	0.534	0.536	0.503	0.397	0.313	0.705	0.556	0.511
50×20	0.462	0.435	0.462	0.483	0.449	0.458	0.483	0.345	0.379	0.645	0.481	0.462
100×5	0.089	0.105	0.101	0.085	0.100	0.098	0.105	0.099	0.102	0.070	0.082	0.095
100×10	0.251	0.255	0.244	0.246	0.241	0.238	0.256	0.278	0.248	0.237	0.234	0.248
100×20	0.473	0.462	0.452	0.473	0.452	0.458	0.463	0.503	0.437	0.512	0.461	0.467
200×10	-0.657	-0.661	-0.663	-0.670	-0.668	-0.665	-0.657	-0.657	-0.666	-0.673	-0.672	-0.665
200×20	-0.846	-0.841	-0.856	-0.851	-0.860	-0.860	-0.837	-0.837	-0.862	-0.849	-0.855	-0.851
500×20	-1.889	-1.888	-1.891	-1.890	-1.891	-1.890	-1.885	-1.888	-1.890	-1.887	-1.891	-1.889
Avg.	-0.115	-0.121	-0.120	-0.117	-0.122	-0.122	-0.116	-0.126	-0.146	-0.090	-0.121	-0.120

We can see that the average AAC results and those of MRSILS(BSCH) are very similar. We applied a Wilcoxon signed-rank test between MRSILS(BSCH) and each one of the algorithms from A_0 to A_9 to verify whether the differences are statistically significant or not. For a significance level of 0.05, with Bonferroni correction, the results indicate that the difference is significant for A_2 ($p < 0.01$) and A_3, A_4, A_7 and A_9 ($p < 0.001$). We also have computed the average solution quality of a random derivation of the grammar over 50 samples and have found it to be $1.04\,\%$. This shows that `irace` is effective in selecting good algorithms.

The algorithm with the best overall ARPD, A_7, is an IGA that alternates between `insertFPR` and `swapInc`. Since the former is similar to the strategy in

MRSILS(BSCH), the observed improvement is probably due to the choice of an IGA and the use of a second, swap-based local search.

It is important to mention a difference regarding the running time of BSCH: while [23] report a steep increase in the average running time for the 3 larger instance groups, we observed a much better scaling. For example, for the instances with 500 jobs and 20 machines we have an average time of 7 s, which is faster by a factor of 30. This is possibly due to a more efficient implementation, so we also evaluate the heuristics on the larger instances of [35].

We conduct an additional experiment to evaluate A_7 on the instances of [35] and compare it to MRSILS(BSCH). The results are presented as the ARPD from MRSILS(BSCH) in Table 3, in which the instances are divided in "Small" and "Large" according to the number of jobs. Five replications per instance were performed for both algorithms. We note an improvement of 0.064% on average for the smaller instances, while the results are very close for the larger ones. This behavior is expected, since the training set in `irace` contained instances that were more similar in dimension to the small or medium-sized instances in this benchmark. Nevertheless, its interesting to see how well the algorithm scales to larger instances. A Wilcoxon signed-rank test confirmed that the difference between the two methods is statistically significant ($p < 2.2 \times 10^{-16}$). It is also worth mentioning that the best algorithm obtained via AAC never performed worse than MRSILS(BSCH),

These results support the observation that we were able to reproduce the state-of-the-art results for the PFSSP with total completion time minimization by means of AAC.

Table 3. ARPD to MRSILS(BSCH) for the VFR benchmark.

Small				Large			
Inst.	ARPD	Inst.	ARPD	Inst.	ARPD	Inst.	ARPD
10×5	0.000	40×5	−0.126	100×20	−0.043	500×20	−0.002
10×10	0.000	40×10	−0.147	100×40	−0.042	500×40	−0.005
10×15	0.000	40×15	−0.111	100×60	−0.020	500×60	−0.003
10×20	0.000	40×20	−0.109	200×20	−0.006	600×20	−0.001
20×5	0.000	50×5	−0.042	200×40	−0.010	600×40	−0.003
20×10	0.000	50×10	−0.182	200×60	−0.013	600×60	−0.004
20×15	0.000	50×15	−0.090	300×20	0.000	700×20	−0.002
20×20	−0.000	50×20	−0.143	300×40	−0.010	700×40	−0.002
30×5	−0.085	60×5	−0.005	300×60	−0.011	700×60	−0.004
30×10	−0.069	60×10	−0.076	400×20	−0.001	800×20	−0.002
30×15	−0.034	60×15	−0.153	400×40	−0.002	800×40	−0.001
30×20	−0.024	60×20	−0.129	400×60	−0.005	800×60	−0.004
Avg.			−0.064				−0.008

4   Conclusions

In this work we propose an automated approach to design heuristic search methods for the PFSSP with total completion time minimization. We implemented algorithmic components found in several works in the literature and propose a context-free grammar which defines how to combine them. We then established a parametric representation and used the automated parameter configuration tool `irace` to search for good algorithms. Our methodology was evaluated on two well-known benchmarks from the literature and the results showed that the obtained algorithms are comparable to the current state of the art with a slightly superior performance.

In summary, this study contributes some evidence that AAC techniques are able to reduce the effort of researchers compared to more manual configuration methods, allowing them to spend more time focusing on creativity-related tasks, such as the design of new components.

As a future work we intend to extend the grammar to include different variants of the PFSSP and different objective functions, possibly integrating them in a single package.

References

1. López-Ibáñez, M., Stützle, T.: The automatic design of multiobjective ant colony optimization algorithms. IEEE Trans. Evol. Comput. **16**(6), 861–875 (2012). https://doi.org/10.1109/TEVC.2011.2182651
2. Burke, E.K., Hyde, M.R., Kendall, G.: Grammatical evolution of local search heuristics. IEEE Trans. Evol. Comput. **16**(3), 406–417 (2012). https://doi.org/10.1109/TEVC.2011.2160401
3. KhudaBukhsh, A.R., Xu, L., Hoos, H.H., Leyton-Brown, K.: Satenstein: automatically building local search SAT solvers from components. Artif. Intell. **232**(Supplement C), 20–42 (2016). https://doi.org/10.1016/j.artint.2015.11.002
4. Branke, J., Nguyen, S., Pickardt, C.W., Zhang, M.: Automated design of production scheduling heuristics: a review. IEEE Trans. Evol. Comput. **20**(1), 110–124 (2016). https://doi.org/10.1109/TEVC.2015.2429314
5. Graham, R., Lawler, E., Lenstra, J., Kan, A.: Optimization and approximation in deterministic sequencing and scheduling: a survey. In: Hammer, P., Johnson, E., Korte, B. (eds.) Discrete Optimization II, Annals of Discrete Mathematics, vol. 5, pp. 287–326. Elsevier (1979). https://doi.org/10.1016/S0167-5060(08)70356-X
6. Garey, M.R., Johnson, D.S., Sethi, R.: The complexity of flowshop and jobshop scheduling. Math. Oper. Res. **1**(2), 117–129 (1976). https://doi.org/10.1287/moor.1.2.117
7. Vázquez-Rodríguez, J.A., Ochoa, G.: On the automatic discovery of variants of the NEH procedure for flow shop scheduling using genetic programming. J. Oper. Res. Soc. **62**(2), 381–396 (2011). https://doi.org/10.1057/jors.2010.132
8. Nawaz, M., Enscore, E.E., Ham, I.: A heuristic algorithm for the m-machine, n-job flow-shop sequencing problem. Omega **11**(1), 91–95 (1983)

9. Mascia, F., López-Ibáñez, M., Dubois-Lacoste, J., Stützle, T.: From grammars to parameters: automatic iterated greedy design for the permutation flow-shop problem with weighted tardiness. In: Nicosia, G., Pardalos, P. (eds.) LION 2013, vol. 7997, pp. 321–334. Springer, New York (2013). https://doi.org/10.1007/978-3-642-44973-4_36

10. Marmion, M.E., Mascia, F., López-Ibáñez, M., Stützle, T.: Automatic design of hybrid stochastic local search algorithms. In: Blesa, M.J., Blum, C., Festa, P., Roli, A., Sampels, M. (eds.) HM 2013, vol. 7919, pp. 144–158. Springer, Heidelberg (2013). https://doi.org/10.1007/978-3-642-38516-2_12

11. Lourenço, N., Pereira, F.B., Costa, E.: Unveiling the properties of structured grammatical evolution. Genetic Program. Evol. Mach. **17**(3), 251–289 (2016). https://doi.org/10.1007/s10710-015-9262-4

12. Hutter, F., Hoos, H.H., Leyton-Brown, K., Stützle, T.: Paramils: an automatic algorithm configuration framework. J. Artif. Intell. Res. **36**, 267–306 (2009)

13. Hutter, F., Hoos, H.H., Leyton-Brown, K.: Sequential model-based optimization for general algorithm configuration. In: Coello, C.A.C. (ed.) LION 5. LNCS, vol. 6683, pp. 507–523. Springer, Heidelberg (2011). https://doi.org/10.1007/978-3-642-25566-3_40

14. López-Ibáñez, M., Dubois-Lacoste, J., Cáceres, L.P., Birattari, M., Stützle, T.: The irace package: iterated racing for automatic algorithm configuration. Oper. Res. Perspect. **3**(Supplement C), 43–58 (2016). https://doi.org/10.1016/j.orp.2016.09.002

15. McKay, R.I., Hoai, N.X., Whigham, P.A., Shan, Y., O'Neill, M.: Grammar-based genetic programming: a survey. Genetic Program. Evol. Mach. **11**(3), 365–396 (2010). https://doi.org/10.1007/s10710-010-9109-y

16. Ruiz, R., Stützle, T.: A simple and effective iterated greedy algorithm for the permutation flowshop scheduling problem. Eur. J. Oper. Res. **177**(3), 2033–2049 (2007). https://doi.org/10.1016/j.ejor.2005.12.009

17. Dong, X., Chen, P., Huang, H., Nowak, M.: A multi-restart iterated local search algorithm for the permutation flow shop problem minimizing total flow time. Comput. Oper. Res. **40**(2), 627–632 (2013). https://doi.org/10.1016/j.cor.2012.08.021

18. Metropolis, N., Rosenbluth, A.W., Rosenbluth, M.N., Teller, A.H., Teller, E.: Equation of state calculations by fast computing machines. J. Chem. Phys. **21**(6), 1087–1092 (1953). https://doi.org/10.1063/1.1699114

19. Dubois-Lacoste, J., Pagnozzi, F., Stützle, T.: An iterated greedy algorithm with optimization of partial solutions for the makespan permutation flowshop problem. Comput. Oper. Res. **81**, 160–166 (2017). https://doi.org/10.1016/j.cor.2016.12.021

20. Framinan, J.M., Leisten, R., Ruiz-Usano, R.: Efficient heuristics for flowshop sequencing with the objectives of makespan and flowtime minimisation. Eur. J. Oper. Res. **141**(3), 559–569 (2002). https://doi.org/10.1016/S0377-2217(01)00278-8

21. Liu, J., Reeves, C.R.: Constructive and composite heuristic solutions to the $P \parallel \sum C_i$ scheduling problem. Eur. J. Oper. Res. **132**(2), 439–452 (2001). https://doi.org/10.1016/S0377-2217(00)00137-5

22. Rad, S.F., Ruiz, R., Boroojerdian, N.: New high performing heuristics for minimizing makespan in permutation flowshops. Omega **37**(2), 331–345 (2009). https://doi.org/10.1016/j.omega.2007.02.002

23. Fernandez-Viagas, V., Framinan, J.M.: A beam-search-based constructive heuristic for the PFSP to minimise total flowtime. Comput. Oper. Res. **81**(Supplement C), 167–177 (2017). https://doi.org/10.1016/j.cor.2016.12.020

24. Benavides, A.J., Ritt, M.: Iterated local search heuristics for minimizing total completion time in permutation and non-permutation flow shops. In: Proceedings of the Twenty-Fifth International Conference on International Conference on Automated Planning and Scheduling, ICAPS 2015, pp. 34–41. AAAI Press (2015)

25. Jarboui, B., Eddaly, M., Siarry, P.: An estimation of distribution algorithm for minimizing the total flowtime in permutation flowshop scheduling problems. Comput. Oper. Res. **36**(9), 2638–2646 (2009). https://doi.org/10.1016/j.cor.2008.11.004

26. Tasgetiren, M.F., Pan, Q.K., Suganthan, P., Chen, A.H.L.: A discrete artificial bee colony algorithm for the total flowtime minimization in permutation flow shops. Inf. Sci. **181**(16), 3459–3475 (2011). https://doi.org/10.1016/j.ins.2011.04.018

27. Rajendran, C., Ziegler, H.: An efficient heuristic for scheduling in a flowshop to minimize total weighted flowtime of jobs. Eur. J. Oper. Res. **103**(1), 129–138 (1997). https://doi.org/10.1016/S0377-2217(96)00273-1

28. Deroussi, L., Gourgand, M., Norre, S.: New effective neighborhoods for the permutation flow shop problem. Technical report LIMOS/RR-06-09, LIMOS/ISIMA (2006)

29. Dubois-Lacoste, J.: Anytime local search for multi-objective combinatorial optimization: design, analysis and automatic configuration. Ph.D. thesis, IRIDIA, Université Libre de Bruxelles, Brussels, Belgium (2014)

30. Kalczynski, P.J., Kamburowski, J.: An improved NEH heuristic to minimize makespan in permutation flow shops. Comput. Oper. Res. **35**(9), 3001–3008 (2008). https://doi.org/10.1016/j.cor.2007.01.020

31. Kalczynski, P.J., Kamburowski, J.: An empirical analysis of the optimality rate of flow shop heuristics. Eur. J. Oper. Res. **198**(1), 93–101 (2009). https://doi.org/10.1016/j.ejor.2008.08.021

32. Vasiljevic, D., Danilovic, M.: Handling ties in heuristics for the permutation flow shop scheduling problem. J. Manuf. Syst. **35**(Supplement C), 1–9 (2015). https://doi.org/10.1016/j.jmsy.2014.11.011

33. Balaprakash, P., Birattari, M., Stützle, T.: Improvement strategies for the F-race algorithm: sampling design and iterative refinement. In: Bartz-Beielstein, T., et al. (eds.) HM 2007, vol. 4771, pp. 108–122. Springer, Heidelberg (2007). https://doi.org/10.1007/978-3-540-75514-2_9

34. Taillard, E.: Benchmarks for basic scheduling problems. Eur. J. Oper. Res. **64**(2), 278–285 (1993). https://doi.org/10.1016/0377-2217(93)90182-M

35. Vallada, E., Ruiz, R., Framinan, J.M.: New hard benchmark for flowshop scheduling problems minimising makespan. Eur. J. Oper. Res. **240**(3), 666–677 (2015). https://doi.org/10.1016/j.ejor.2014.07.033

36. Pan, Q.K., Ruiz, R.: Local search methods for the flowshop scheduling problem with flowtime minimization. Eur. J. Oper. Res. **222**(1), 31–43 (2012). https://doi.org/10.1016/j.ejor.2012.04.034

Data Clustering Using Grouping Hyper-heuristics

Anas Elhag$^{(\boxtimes)}$ and Ender Özcan

ASAP Research Group, School of Computer Science,
University of Nottingham, Nottingham, UK
Anas.Abdalla@outlook.com, Ender.Ozcan@nottingham.ac.uk

Abstract. Grouping problems represent a class of computationally hard to solve problems requiring optimal partitioning of a given set of items with respect to multiple criteria varying dependent on the domain. A recent work proposed a general-purpose selection hyper-heuristic search framework with reusable components, designed for rapid development of grouping hyper-heuristics to solve grouping problems. The framework was tested only on the graph colouring problem domain. Extending the previous work, this study compares the performance of selection hyper-heuristics implemented using the framework, pairing up various heuristic/operator selection and move acceptance methods for data clustering. The selection hyper-heuristic performs the search processing a single solution at any decision point and controls a fixed set of generic low level heuristics specifically designed for the grouping problems based on a bi-objective formulation. An archive of high quality solutions, capturing the trade-off between the number of clusters and overall error of clustering, is maintained during the search process. The empirical results verify the effectiveness of a successful selection hyper-heuristic, winner of a recent hyper-heuristic challenge for data clustering on a set of benchmark problem instances.

Keywords: Heuristic · Multiobjective optimisation
Reinforcement learning · Adaptive move acceptance

1 Introduction

Solving a grouping problem which is an NP-hard combinatorial optimisation problem [1] requires partitioning of set of objects/items into a minimal collection of mutually disjoint subsets. Different grouping problems impose different problem specific constraints, and introduce different objectives to optimise. Consequently, not all groupings are allowed for all problems, since a solution must satisfy the problem specific constraints. These problem specific constraints/objectives forbid the 'trivial' solutions which consist of placing all the objects into one group.

© Springer International Publishing AG, part of Springer Nature 2018
A. Liefooghe and M. López-Ibáñez (Eds.): EvoCOP 2018, LNCS 10782, pp. 101–115, 2018.
https://doi.org/10.1007/978-3-319-77449-7_7

Data clustering is a grouping problem which requires partitioning a given set of data items or property vectors into a minimal number of disjoint clusters/groups, such that the items in each group are *close* to each other with respect to a given similarity measure, and are *distant* from the items in the other groups with respect to the same measure. Data clustering plays an important role in many disciplines where there is a need to learn the inherent grouping structure of data such as data mining and bioinformatics [2–4].

Selection hyper-heuristics have emerged as metaheuristics searching the space formed by operators/heuristics for solving combinatorial optimisation problems [5]. In this study, we use a bi-objective formulation of data clustering as a grouping problem with the goal of simultaneously optimising the number of clusters and error/cost. [6] proposed a grouping hyper-heuristic framework for solving grouping problems, exploiting the bi-objective nature of the grouping problems to capture the best of the two worlds. The results showed the potential of selection hyper-heuristics based on the framework for graph colouring. In here, we extend that previous work, apply and compare the performance of selection hyper-heuristics implemented based on the grouping hyper-heuristic framework for data clustering on a set of well-known benchmarks.

Section 2 discusses commonly used encoding schemes to represent solutions to grouping problems and related work on data clustering. Section 3 describes the selection hyper-heuristic framework for grouping problems. The details of the experimental design, including the benchmark instances, tested selection hyper-heuristics, parameter settings and evaluation criteria used for performance comparison of algorithms, are given in Sect. 4. Section 4.4 presents the empirical results and finally, conclusion is provided in Sect. 5.

2   Related Work

2.1   Solution Representation in Grouping Problems

A variety of encoding schemes and population-based operators have been proposed and applied to various grouping problems [1]. Examples include the Numeric Encoding (NE) [7] which uses a constant-length encoding in which each position corresponds to one object; the Locus-Based Adjacency (LBA) representation [4,8] and the Linear Linkage Encoding (LLE) [9], which are linkage-based fixed-length encodings in which each location represents one object and stores an integer value that represents a link to *another* object in the same group. However, most of these schemes violate one or more of the six design principles for constructing useful representations [10]. For instance, the NE and LBA representations allow multiple chromosomes in the search space to represent the same solution [11], and hence violate the "minimal redundancy" principle, which states that each solution should be represented by as few distinct points in that search space as possible; ideally one point only.

In this study we implemented a modified version of the Grouping Genetic Algorithm Encoding (GGAE) representation which was proposed as part of a genetic algorithm that was heavily modified to suit the structure of the grouping problems [12], and has successfully been applied in many real world problems such as data clustering [2] and machine-part cell formation [13]. The encoding in the GGAE consists of two parts. The object membership part is used only to identify which objects are in which groups. No operators are applied to this part. The groups part encodes the groups on a 'one gene per group' basis. The group part is written after the standard NE part, and the two are separated by a colon. For example, a solution that puts the first three items in one group and the last two items in a different group is represented as follows: $\{D, D, D, C, C : D, C\}$. GGAE operators are applied only to the groups part, and might lead to increasing or decreasing the number of groups in the given solution. This approach implies that the length of the GGAE encoding is not fixed and can not be known in advance. Further discussion of grouping representations could be found in [1,6].

2.2   Data Clustering

Given a set X of n vectors in a given feature space S, $X=\{x_1, x_2, x_3, \ldots, x_n\}$, the data clustering problem requires finding an optimal partition $U=\{u_1, u_2, u_3, \ldots, u_k\}$ of X, where u_i is the ith cluster/group of U, such that an overall similarity measure between the vectors that belong to the same group, or an overall dissimilarity measure of the vectors that belong to different groups, is maximized, in terms of a given cost function $f(U)$.

There are different supervised and unsupervised measures that are used to evaluate the clustering results [14], including the Sum of Quadratic Errors (SSE) [2], the Silhouette Width (SW) [15], the Davies-Bouldin Index (DB) [16] and the Rand Index (R) [17], among others. The most well-known evaluation method in the data clustering literature is based on the Euclidean distance which is used to give an overall measure of the error of the clustering. In this study, a well-known unsupervised distance measure known as the 'sum of quadratic errors' (SSE) is adopted. Assuming that each data item has p properties, the SSE calculates a centroid vector for each cluster. The resulting centroid is also composed of p values, one for each property. The sum of the distances; i.e. errors; of each item's properties corresponds to the distances to the property values of the centroid of the cluster to which the item belongs.

$$error = \sum_{l=1}^{k} \sum_{i=1}^{n} W_{il} \sum_{j=1}^{p} (x_{ij} - u_{lj})^2 \tag{1}$$

$k \equiv$ the number of clusters, $n \equiv$ the number of items, $p \equiv$ the number of properties, $W_{il} = 1$ if the ith item is in k^{th} cluster, and 0 otherwise, $x_{ij} \equiv$ the ith item's j^{th} property, and $u_{lj} \equiv$ the center of the jth property of the kth cluster. Traditionally, clustering approaches are classified to partitional, hierarchical and density-based approaches [18]. [8,19] categorize the clustering approaches into 3 and 4 groups, respectively, based on the clustering criterion being optimized. The first group in both studies consists of the clustering algorithms that look for compact clusters by optimizing intra-cluster variations, such as variations between items that belong to the same cluster or between the items and cluster representatives. Well-known algorithms such as the k-means [20], model-based clustering [21], average link agglomerative clustering [22] and self-organizing maps [23] belong to this category. The second group, consists of the clustering algorithms that strive for connected clusters by grouping neighboring data into the same cluster. Classical clustering techniques such as single link agglomerative clustering [22] and density-based methods [24] belong to this group.

The third group according to [8] consists of clustering algorithms that look for spatially-separated clusters. However, this objective on its own may result in clustering the oultiers individually while merging the rest of the data items into one big cluster. Clustering objective such as the Dunn Index and the Davies-Bouldin Index [14] combine this objective with other clustering objectives, such as compactness and connectedness, in order to improve the resulting solution. In contrast to the first two groups, this objective has not been used in any specialized clustering algorithm. According to [19], the third group includes simultaneous row-column clustering techniques known as bi-clustering algorithms [25]. Finally, the fourth group according to [19] includes the multi-objective clustering algorithms that seek to optimize different characteristics of the given data set [3] along with the clustering ensembles approaches [26].

3   A Grouping Hyper-heuristic Approach to Solve Grouping Problems

Algorithm 1 describes the steps of the grouping approach used in this study. Firstly, an initial set of $(UB - LB + 1)$ non-dominated solutions is generated such that there is exactly 1 solution for each value of $k \in [LB, UB]$ (steps 1–3 of Algorithm 1). Consequently, one of the low level heuristics is selected (using the hyper-heuristic selection method) and applied on a solution that is randomly selected from the current set of solutions (steps 5–8 of Algorithm 1). It is vital to ensure that none of the solutions in the non-dominated set that is maintained by the framework break the dominance requirement throughout the search process. To this end, an adaptive acceptance mechanism that involves multiple tests is introduced. Traditionally, the hyper-heuristic's move acceptance makes the final decision regarding the acceptance of a solution. In our approach however, this component acts only as a pre-test for the final acceptance. New solutions are accepted only after successfully passing multiple tests. The decision made by the traditional hyper-heuristic move acceptance only indicates whether the new solution s_{new} is to be considered for acceptance by the grouping framework (step 9 of Algorithm 1). s_{new} could still be a worsening solution based on the nature of the traditional move acceptance used. At this point, two main possibilities, one of which involves the application of local search to further improve the set of non-dominated solutions:

1. If a worsening solution s_{new} passes the traditional move acceptance pre-test, it does not immediately replace the solution s_i in the non-dominated

set (steps 10 of Algorithm 1). Instead, s_{new} is compared to s_{i-1}. Only if the cost value of s_{new} is better than s_{i-1}, it gets to replace s_i in the non-dominated set. Otherwise, s_{new} is rejected, despite having passed the pre-test (steps 11–16 of Algorithm 1).

2. If an improving solution s_{new} passes the traditional move acceptance pre-test, it replaces the solution s_i in the non-dominated set immediately without further tests (step 17 of Algorithm 1). However, this replacement might lead to a violation of the dominance rule if s_{new}, which has fewer groups than s_{i+1}, also has a better cost value than s_{i+1} (step 2 of Algorithm 2). This situation leads to two cases:

2.1 If the cost value of s_{i+1} is better than that of s_{new}, then no violations have occurred to the dominance rule, and no further action is required (steps 2–3 of Algorithm 2).

2.2 If the cost value of s_{i+1} is worse than that of s_{new}, then s_{i+1} violates the dominance rule and hence the framework removes it from the set of non-dominated solutions being maintained. The framework then applies one of the divide heuristics on s_{new} in order to generate a new solution to replace the solution that has been removed (steps 4–8 of Algorithm 2). The *for* loop in Algorithm 2 is to repeat this process for solutions at $i+1$ and $i+2$. In the worst case scenario, all the solutions between i and UB will get replaced. This process can be considered as local search.

3.1   Low Level Heuristics

Three types of low level heuristics were implemented in this study. *Merge Heuristics* merge two groups, u_i and u_j, into one, u_l, and decrease the number of grouping in the selected solution. 3 merge heuristics were developed. In **M1**, the 2 groups to be merged are selected at random. In **M2**, the 2 groups containing the fewest items are merged. In **M3**, the 2 groups with the smallest partial costs are merged. These heuristics yields big jumps in the search space, and hence, are regarded to be diversifying components.

Similarly, *Divide Heuristics* divide a selected group u_i into two, u_{i1} and u_{i2}, and increase the number of groups in the given solution. 3 versions of the divide heuristic were developed. In **D1**, a group that is selected at random is divided. In **D2**, the group to be divided is the group with the most number of items. In **D3**, the group with the biggest partial cost value is divided. These heuristics can be considered as intensifying components.

Change Heuristics attempt to make small alterations in a selected solution by moving selected items between different groups while preserving the original number of the groupings in the selected solution. 4 change heuristics have been developed. In **C1**, a randomly selected item is moved to a randomly selected group. In **C2**, the item with the largest number of conflicts in a group is moved into a randomly selected group. **C3** and **C4** find the item with the largest number of conflicts in the group with the largest number of conflicts. C3 moves this item into a randomly selected group, while C4 moves it into the group with the minimum number of conflicts.

Algorithm 1. A Grouping Hyper-heuristic Framework

1: Create an initial set of non-dominated solutions, containing 1 solution for each value of $k \in [LB, UB]$.
2: Calculate the cost values of all solutions in the solutions set.
3: Keep an external archive copy of the solutions set in order to keep track of the best solutions found.
4: **while** (elapsedTime < maxTime) **do**
5:    Randomly select a solution s_j from the current set of non-dominated solutions $j \leftarrow UniformRandom(LB, UB)$.
6:    Select one of the low level heuristics, LLH.
7:    $s_{new} \leftarrow Apply(LLH, s_j)$ {s_{new} contains $i = (j-1)$ or j or $(j+1)$ groups based on the applied LLH}.
8:    Calculate $f(s_{new})$.
9:    $result \leftarrow moveAcceptance(s_{new}, s_i)$. {// Compare the cost value of s_{new} to the cost value of s_i from the current non-dominated set using the move acceptance method returning $ACCEPT$ or $REJECT$}
   {// When a worsening solution passes the traditional *moveAcceptance* pre-test, it is handled as follows}
10:    **if** (($result$ is $ACCEPT$) **and** ($f(s_{new}) > f(s_i)$)) **then**
11:      **if** ($f(s_{new}) > f(s_{i-1})$) **then**
12:        Do nothing. {// s_{new} is rejected}
13:      **else**
14:        $s_i \leftarrow s_{new}$. {// s_i is replaced by s_{new} in the non-dominated set}
15:      **end if**
16:    **end if**
17:    **if** (($result$ is $ACCEPT$) **and** ($f(s_{new}) \leq f(s_i)$)) **then**
18:      $s_i \leftarrow s_{new}$. {// s_i is replaced by s_{new} in the non-dominated set}
19:      $improveNonDominatedSet(i)$
20:    **end if**
   {// if $result$ is $REJECT$ continue}
21: **end while**

Algorithm 2. *ImproveNonDominatedSet(i)*: Aims at improving the cost of solutions in the non-dominated set starting from the ith solution to the UB^{th} using a *divide* heuristic

---

1: **for** ($j = i, UB$) **do**
2:    **if** ($f(s_{(j+1)}) \leq f(s_j)$) **then**
3:      $BREAK$. {// No further improvement is possible}
4:    **else**
5:      Randomly select a divide heuristic, $LLDH$.
6:      $s_{new} \leftarrow Apply(LLDH, s_j)$.
7:      $s_{(j+1)} \leftarrow s_{new}$. {// $s_{(j+1)}$ is replaced by s_{new} in the non-dominated set}
8:    **end if**
9: **end for**

3.2 Selection Hyper-heuristic Components

An investigation of the performance of the grouping hyper-heuristic framework over a set of selected benchmark instances from the data clustering problem domain is carried out using different selection hyper-heuristic implementations. A total of 9 selection hyper-heuristics is generated using the combinations of the Simple Random (SR), Reinforcement Learning (RL) and Adaptive Dynamic Heuristic Set (ADHS) heuristic selection methods, and the Late Acceptance (LACC), Great Deluge (GDEL) and Iteration Limited Threshold Accepting (ILTA) move acceptance methods. From this point onward, a selection hyper-heuristic will be denoted as *heuristics selection-move acceptance*. For example, SR-GDEL is the hyper-heuristic that combines simple random selection method with great deluge move acceptance criterion.

SR chooses a low level heuristic at random. RL [27] maintains a utility score for each low level heuristic. If a selected heuristic generates an improved solution then its score is increased by one, otherwise it is decreased by one. At each decision point, the heuristic with the maximum score is selected. LACC [28] accepts all improving moves, however a worsening current solution is compared to a previous solution which was visited at a fixed number of steps prior during the search. If the current solution's objective value is better than that previous solution's objective value, it is accepted. Hence, a fixed size list containing previous solutions is maintained and this list gets updated at each step. A slightly modified version of GDEL is used in here and multiple lists are maintained, where a list is formed for each active group number (number of clusters).

GDEL [29] sets a target objective value and accepts all solutions whether improving or worsening as long as the objective value of the current solution is better than the target value. The target values is often taken as the objective value of the initial solution and it is decreased linearly in time towards a minimum expected objective value τ_t as shown in Eq. 2.

$$\tau_t = f_0 + \Delta F \times (1 - \frac{t}{T}) \qquad (2)$$

In this equation, T is the maximum number of steps, ΔF is an expected range for the maximum objective value change, and f_0 is the final objective value. ADHS-ILTA [30] is one of the best performing hyper-heuristics in the literature. This elaborate online learning hyper-heuristic won the CHeSC 2011 competition across six hard computational problem domains [31]. The learning heuristic selection method consists of various mechanisms, such as for creating new heuristics vi relay hybridisation or for excluding the low level heuristics with poor performance. The move acceptance component is adaptive threshold method. The readers can refer to [30] for more details on this hyper-heuristic.

4 Application of Grouping Hyper-heuristics to Data Clustering

In this section, we provide the performance comparison of nine selection hyper-heuristics formed by the combination of {SR, RL, DH} heuristic selection and

{ILTA, LACC, GDEL} move acceptance methods for data clustering. The performance of the approaches proposed based on the developed framework are further compared to previous approaches from the literature.

4.1 Experimental Data

16 data clustering problem instances that have different properties and sizes were used in this study. The first 12 of these instances, shown in Table 1 and Fig. 1, were taken from [8]. The top 6 instances are 2-D hand-crafted problem instances that contain interesting data properties, such as different cluster sizes and high degrees of overlap between the clusters. Instances Square1, Square4 and Sizes5 contain four clusters each. The main difference between these instances is that clusters in Square1 and Square4 are of equal size, whereas the clusters in Sizes5 are not. On the other hand, data instance Long1 consists of two well-connected long clusters [8,19]. Problem instances Twenty and Forty exhibit a mixture of the properties discusses above. The 6 problem instances on the bottom half of Fig. 1 are randomly generated instances that were created using the Gaussian cluster generator described in [8]. Instances 2D-4c, 2D-10c, 2D-20c and 2D-40c are all 2 dimensional instances containing 4, 10, 20 and 40 clusters respectively. Similarly,

Table 1. The characteristics of the synthetic, Gaussian and real-world data clustering problem instances used during the experiments. N is the number of items, D is the number of dimensions/attributes and k^* is the best number of clusters [8]. L and U are the lower and upper bounds for the k values used during the experiments.

	Data clustering				Range (k)	
	Instance	N	D	k^*	LB	UB
Synthetic	Square1	1000	2	4	2	9
	Square4	1000	2	4	2	9
	Sizes5	1000	2	4	2	9
	Long1	1000	2	4	2	9
	Twenty	1000	2	20	16	24
	Fourty	1000	2	40	36	44
Gaussian	2D-4c	1623	2	4	2	9
	2D-10c	2525	2	10	6	14
	2D-20c	1517	2	20	16	24
	2D-40c	2563	2	40	36	44
	10D-4c	958	10	4	2	9
	10D-10c	3565	10	10	6	14
Real	Zoo	101	16	7	3	11
	Iris	150	4	3	2	7
	Dermatology	366	34	6	2	10
	Breast-cancer	699	9	2	2	7

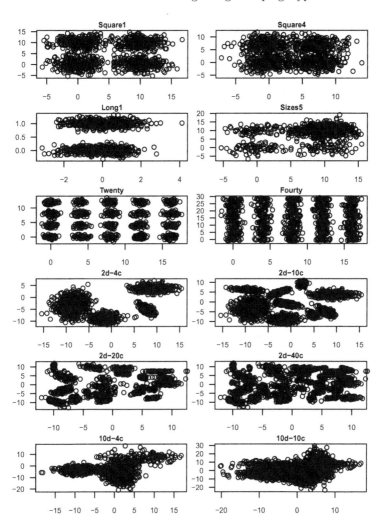

Fig. 1. The synthetic data clustering problem instances used in the experiments.

instances 10D-4c and 10D-10c are 10 dimensional instances that contain 4 and 10 clusters respectively.

Additionally, 4 real world problem instances were used in this study. These are taken from the UCI Machine Learning Repository [32], which maintains multiple data sets as a service to the machine learning community. These selected real-world instances differ from each other in many ways, such as in the number of clusters, number of dimensions as well as the data type of the values in each dimension. For example, Iris instance contains 3 equal-sized clusters of 50 data items each, and the data type of the values of each dimension is continuous, whereas Dermatology instance contains 6 clusters of different sizes of $(112, 61, 72, 49, 52, 20)$ data items, and the data type of each dimension is integer.

4.2 Trials and Parameters Settings and CPU Specifications

Based on initial experiments, the initial score value of all the LLHs in the RL selection method were set to (upper score bound - 2 * number of heuristics). The upper and lower score bounds of each one of the LLHs are set to 40 and 0 respectively, and the score increments and decrements values are both set to 1. The GDEL parameters are set to the following values: T is set to the maximum duration of a trial, ΔF is set to the minimum cost value in the initial non-dominated set and f_0 is set to 0 [29]. A LACC approach that uses k separate lists of equal lengths, one for each value of $k \in [LB, UB]$, is adopted based on the results of some initial experiments. A separate list of 50 previous solutions is maintained for each one of the solutions in the current set of non-dominated solutions. The parameters of the ADHS and ILTA followed the same settings as suggested in the literature [30].

30 initial solutions are created randomly for each one of the problem instances. In order to avoid initialization bias, all hyper-heuristic approaches operated on the same 30 initial solutions for each problem instance. Each experiment was repeated 30 times, for 600 s each. Experiments were conducted on 3.6 GHz Intel Core $i7 - 3820$ machines with 16.0 GB of memory, running on "Windows 7 OS".

4.3 Evaluation Criteria

In each experiment, each one of the 9 different hyper-heuristics used in this study starts from an initial set of non-dominated solutions and tries to find the best non-dominated set that can be found in the given period of time allowed for each run. The overall success of a hyper-heuristic is evaluated using *success rate*, denoted as $(sRate\%)$ which indicates the percentage of the runs in which the expected (best) number of groups/corner points has been successfully found by a given algorithm; and the average time (in seconds) taken to achieve those success rates which is calculated using the duration of the successful runs only.

4.4 Experimental Results and Remarks

The actual values for the different dimensions within each one of the instances described above were measured on different scales. Consequently, some data processing was carried out before applying the grouping hyper-heuristics on these clustering instances. In this pre-processing, the real data is normalized such that the mean is equal to 0 and the standard deviation is equal to 1 in each dimension.

Initial experiments were conducted to observe the behavior of the grouping hyper-heuristic framework considering different k values for each problem instance, as specified in Table 1, while the heuristic selection is fixed from {SR,

RL, DH} and the heuristic acceptance is fixed from {ILTA, LACC, GDEL}. A thorough performance analysis of the hyper-heuristics is performed. Then the performance of the hyper-heuristic with the best mean corner point is compared to the performance of some previously proposed approaches.

The results are summarised in Tables 2, 3 and 4, showing the average best grouping, the standard deviation and the success rate for each of the grouping hyper-heuristics for each of the instances over 30 runs.

Table 2. The Performance of Reinforcement Learning Selection Hyper-heuristics: the success rate ($sRate\%$), the average best number of clusters ($\mu(k_{best})$) and the standard deviation ($\sigma(k_{best})$) of each hyper-heuristic approach on the data clustering problem instances over the 30 runs.

	Instance	k^*	RL-ILTA			RL-LACC			RL-GDEL		
			sRate%	$\mu(k_{best})$	$\sigma(k_{best})$	sRate%	$\mu(k_{best})$	$\sigma(k_{best})$	sRate%	$\mu(k_{best})$	$\sigma(k_{best})$
Synthetic	Square1	4	100.00	4.0	±0.0	86.67	4.13	±0.35	93.33	4.07	±0.25
	Square4	4	96.67	4.03	±0.18	93.33	4.07	±0.25	86.67	4.13	±0.35
	Sizes5	4	83.33	3.9	±0.40	70.00	4.03	±0.56	73.33	4.03	±0.61
	Long1	4	13.33	5.53	±2.19	3.33	6.07	±2.07	3.33	5.83	±1.98
	Twenty	20	23.33	20.87	±1.36	16.67	20.9	±1.37	6.67	20.87	±1.38
	Fourty	40	20.00	41.07	±1.72	16.67	41.43	±1.73	10.00	41.33	±2.04
Gaussian	2D-4c	4	16.67	3.83	±0.59	6.67	3.90	±0.96	10.00	3.83	±0.83
	2D-10c	10	13.33	8.83	±1.21	3.33	8.97	±1.27	10.00	9.17	±1.46
	2D-20c	20	63.33	18.0	±2.30	40.00	18.23	±2.33	50.00	18.73	±2.35
	2D-40c	40	20.00	32.6	±3.24	10.00	32.4	±3.41	13.33	32.77	±3.21
	10D-4c	4	13.33	3.63	±1.07	6.67	3.77	±1.43	3.33	4.13	±1.72
	10D-10c	10	0.00	8.57	±1.74	0.00	8.5	±1.59	0.00	8.9	±1.86
Real	Zoo	7	36.67	7.30	±1.49	13.33	7.6	±1.61	16.67	7.67	±1.60
	Iris	3	20.00	4.47	±1.83	6.67	4.83	±1.84	3.33	4.90	±1.92
	Dermatology	6	20.00	6.37	±1.50	10.00	6.37	±1.52	3.33	6.57	±1.77
	Breast-cancer	2	16.67	3.33	±0.92	6.67	3.43	±0.90	6.67	3.43	±0.82
	Wins		15	9	–	0	3	–	0	7	–

The bottom row of each table, titled 'wins', shows the number of times each grouping selection hyper-heuristic has achieved the best average grouping including the ties with the other algorithms. A standard deviation of ±0.0 for a particular algorithm indicates that this algorithm succeeded to find the best grouping for the particular instance over the 30 runs and achieved a 100% success rate.

In general, hyper-heuristics using the ILTA selection method performs better than the others. Grouping hyper-heuristics which use LACC or GDEL as acceptance method achieved low success rates in most of the instances, scoring a success rate that is less than 40% in most of the real and Gaussian instances. ADHS-ILTA hyper-heuristic receives the most number of wins across all the tested approaches, while RL-ILTA and SR-ILTA follow it in that order, respectively, as could be seen in the tables. ADHS-ILTA delivers the best success rate on all Gaussian instances, and most of the remaining instances. On average, ADHS-ILTA performs better than the other grouping hyper-heuristics on 50%

Table 3. The Performance of Adaptive Dynamic Heuristics Set (ADHS) Selection Hyper-heuristics: the success rate ($sRate\%$), the average best number of clusters ($\mu(k_{best})$) and the standard deviation ($\sigma(k_{best})$) of each hyper-heuristic approach on the data clustering problem instances over the 30 runs.

Instance		k^*	ADHS-ILTA			ADHS-LACC			ADHS-GDEL		
			sRate%	$\mu(k_{best})$	$\sigma(k_{best})$	sRate%	$\mu(k_{best})$	$\sigma(k_{best})$	sRate%	$\mu(k_{best})$	$\sigma(k_{best})$
Synthetic	Square1	4	100.00	4.0	±0.0	83.33	4.23	±0.57	100.00	4.0	±0.0
	Square4	4	100.00	4.0	±0.0	76.67	4.33	±0.66	93.33	4.07	±0.25
	Sizes5	4	80.00	3.87	±0.43	70.00	4.0	±0.64	73.33	4.13	±0.68
	Long1	4	16.67	5.60	±2.22	3.33	6.0	±2.05	3.33	6.27	±2.24
	Twenty	20	36.67	20.93	±1.50	30.00	20.90	±1.56	30.00	21.0	±1.49
	Fourty	40	23.33	41.03	±1.69	6.67	41.6	±1.90	3.33	41.6	±2.08
Gaussian	2D-4c	4	23.33	3.87	±0.51	13.33	4.07	±0.94	6.67	4.3	±1.39
	2D-10c	10	20.00	8.97	±1.22	3.33	8.77	±1.43	10.00	8.9	±1.35
	2D-20c	20	73.33	18.23	±2.24	43.33	18.57	±2.21	23.33	18.63	±2.58
	2D-40c	40	26.67	33.6	±3.80	10.00	32.57	±3.33	6.67	32.63	±3.37
	10D-4c	4	20.00	3.8	±0.961	6.67	4.03	±1.35	6.67	4.17	±1.84
	10D-10c	10	10.00	8.83	±1.62	0.00	8.93	±1.87	0.00	9.1	±1.90
Real	Zoo	7	56.67	7.30	±1.47	36.67	7.37	±1.63	3.33	7.73	±1.72
	Iris	3	26.67	4.5	±1.68	3.33	5.23	±1.45	3.33	5.3	±1.32
	Dermatology	6	26.67	6.37	±1.45	10.00	6.6	±1.57	6.67	6.43	±1.74
	Breast-cancer	2	13.33	3.30	±0.95	3.33	3.67	±0.99	0.00	3.70	±1.06
Wins			16	10	–	0	4	–	0	2	–

Table 4. The Performance of Simple Random (SR) Selection Hyper-heuristics: the success rate ($sRate\%$), the average best number of clusters ($\mu(k_{best})$) and the standard deviation ($\sigma(k_{best})$) of each hyper-heuristic approach on the data clustering problem instances over the 30 runs.

Instance		k^*	SR-ILTA			SR-LACC			SR-GDEL		
			sRate%	$\mu(k_{best})$	$\sigma(k_{best})$	sRate%	$\mu(k_{best})$	$\sigma(k_{best})$	sRate%	$\mu(k_{best})$	$\sigma(k_{best})$
Synthetic	Square1	4	100.00	4.0	±0.0	80.00	4.27	±0.58	86.66	4.17	±0.46
	Square4	4	86.67	4.13	±0.35	83.33	4.3	±.75	73.33	4.57	±1.10
	Sizes5	4	80.00	4.13	±0.63	63.33	4.33	±0.76	70.00	4.13	±0.78
	Long1	4	6.67	5.73	±2.24	6.67	5.90	±2.16	6.67	5.93	±2.35
	Twenty	20	26.67	20.93	±1.39	10.00	21.37	±1.69	16.67	21.27	±1.62
	Fourty	40	3.33	41.27	±1.70	3.33	41.67	±1.94	0.00	41.47	±1.82
Gaussian	2D-4c	4	16.67	4.03	±0.93	3.33	4.3	±1.34	6.67	4.43	±1.38
	2D-10c	10	10.00	8.87	±1.28	3.33	8.73	±1.17	6.67	9.03	±1.30
	2D-20c	20	46.67	18.77	±2.47	26.67	18.93	±2.49	6.67	19.3	±2.47
	2D-40c	40	23.33	33.6	±3.71	3.33	32.6	±3.39	6.67	32.77	±3.51
	10D-4c	4	3.33	3.67	±0.88	0.00	3.6	±1.30	10.00	3.83	±1.49
	10D-10c	10	0.00	8.77	±1.72	0.00	8.7	±1.73	0.00	9.03	±1.73
Real	Zoo	7	36.67	7.33	±1.56	10.00	7.53	±1.55	6.67	7.90	±1.67
	Iris	3	20.00	5.23	±1.41	6.67	5.4	±1.28	10.00	5.47	±1.25
	Dermatology	6	13.33	6.60	±1.61	3.33	6.73	±1.78	3.33	6.73	±1.76
	Breast-cancer	2	6.67	4.03	±1.45	0.00	4.3	±1.62	0.00	4.27	±1.60
Wins			14	12	–	2	0	–	2	5	–

of instances and the standard deviation associated with the average values is the lowest in most of the cases. The performance of ADHS-ILTA is, therefore, taken for further comparison with other known clustering algorithms as shown in Table 5 including "k-means" [20], "Mock" [8], "Ensemble" [26], "Av. Link" [22], and "S. Link" [22]. The comparison results shown in Table 5 are based on the average number of clusters. The 'wins' row at the bottom of the table indicates the number of winning times each approach achieved over all the Gaussian and the Synthetic instances. Given this table, it is observed that the performance of ADHS-ILTA lies on the top of the other approaches equally with MOCK algorithm scoring five wins, and outperforming the other competing algorithms. ADHS-ILTA scored the best average distinctively in four instances including, Twenty, Forty, 2D-4c and 2D-20c.

Table 5. Comparing the performances of different approaches on data clustering problem instances based on the average best number of clusters. The entries in bold indicate the best result obtained by the associated algorithm for the given instance.

	Instance	k^*	ADHS-ILTA	k-means	MOCK	Ensemble	Av. link	S. link
Synthetic	Square1	4	**4.00**	**4.00**	**4.00**	**4.00**	**4.00**	2.72
	Square4	4	**4.00**	**4.00**	**4.00**	4.04	4.26	2.00
	Sizes5	4	**3.87**	3.74	**3.87**	3.70	3.76	2.44
	Long1	4	5.60	8.32	8.34	**4.92**	7.78	2.02
	Twenty	20	20.93	–	–	–	–	–
	Fourty	40	41.03	–	–	–	–	–
Gaussian	2D-4c	4	**3.87**	3.69	3.70	2.23	4.50	4.40
	2D-10c	10	8.97	10.66	**9.65**	4.50	15.20	8.00
	2D-20c	20	**18.23**	21.87	17.31	1.24	16.30	16.3
	2D-40c	40	33.60	30.63	**35.26**	2.27	30.90	27.7
	10D-4c	4	3.80	3.59	3.60	5.37	**4.00**	2.00
	10D-10c	10	8.83	**9.02**	8.88	3.86	5.30	2.00
	Wins		5	3	5	2	2	0

5   Conclusion

In this study, the grouping hyper-heuristic framework, previously applied to graph colouring, is extended to handle the data clustering problem. The performances of various selection hyper-heuristics are compared using a set of benchmark instances which vary in terms of the number of items, groups as well as number and nature of dimensions. This investigation is carried out using different pairwise combinations of the 'simple random', the 'reinforcement learning' and the 'adaptive dynamic heuristic set' heuristic selection methods and the 'late acceptance', 'great deluge' and 'iteration limited threshold accepting' move acceptance methods. The genetic grouping algorithm encoding is used as a solution representation. The best heuristic selection and move acceptance turns out to be the Adaptive Dynamic Heuristic Set and Iteration Limited Threshold Accepting methods. This selection hyper-heuristic, winner of a hyper-heuristic challenge performing well across six different problem domains, is sufficiently general and very effective considering that it still ranks the best algorithm for data clustering as well.

The empirical results show that the proposed framework is indeed sufficiently general and reusable. Also, although the ultimate goal of the grouping framework is not to beat the state of the art techniques that are designed and tuned for specific problems, the results obtained by learning-based hyper-heuristics which uses feedback during the search process turned out to be very competitive when compared to previous approaches from the literature.

References

1. Falkenauer, E.: Genetic Algorithms and Grouping Problems. Wiley, New York (1998)
2. Agustın-Blas, L.E., Salcedo-Sanz, S., Jiménez-Fernández, S., Carro-Calvo, L., Del Ser, J., Portilla-Figueras, J.A.: A new grouping genetic algorithm for clustering problems. Expert Syst. App. **39**(10), 9695–9703 (2012)
3. Mitra, S., Banka, H.: Multi-objective evolutionary biclustering of gene expression data. Pattern Recogn. **39**(12), 2464–2477 (2006)
4. Park, Y.J., Song, M.S.: A genetic algorithm for clustering problems. In: Koza, J.R., Banzhaf, W., Chellapilla, K., Deb, K., Dorigo, M., Fogel, D.B., Garzon, M.H., Goldberg, D.E., Iba, H., Riolo, R. (eds.) Genetic Programming 1998: Proceedings of the Third Annual Conference, University of Wisconsin, Madison, Wisconsin, USA, 22–25 July, pp. 568–575. Morgan Kaufmann (1998)
5. Burke, E.K., Gendreau, M., Hyde, M.R., Kendall, G., Ochoa, G., Özcan, E., Qu, R.: Hyper-heuristics: a survey of the state of the art. JORS **64**(12), 1695–1724 (2013)
6. Elhag, A., Özcan, E.: A grouping hyper-heuristic framework: application on graph colouring. Expert Syst. App. **42**(13), 5491–5507 (2015)
7. Talbi, E.G., Bessiere, P.: A parallel genetic algorithm for the graph partitioning problem. In: Proceedings of the 5th International Conference on Supercomputing, pp. 312–320. ACM (1991)
8. Handl, J., Knowles, J.D.: An evolutionary approach to multiobjective clustering. IEEE Trans. Evol. Comput. **11**(1), 56–76 (2007)
9. Ülker, Ö., Özcan, E., Korkmaz, E.E.: Linear linkage encoding in grouping problems: applications on graph coloring and timetabling. In: Burke, E.K., Rudová, H. (eds.) PATAT 2006. LNCS, vol. 3867, pp. 347–363. Springer, Heidelberg (2007). https://doi.org/10.1007/978-3-540-77345-0_22
10. Radcliffe, N.J.: Formal analysis and random respectful recombination. In: Proceedings of the 4th International Conference on Genetic Algorithm, pp. 222–229 (1991)
11. Radcliffe, N.J., Surry, P.D.: Fitness variance of formae and performance prediction. In: Whitley, L.D., Vose, M.D. (eds.) FOGA, pp. 51–72. Morgan Kaufmann Publishers Inc. (1994)

12. Falkenauer, E.: The grouping genetic algorithms: widening the scope of the GAs. Belg. J. Oper. Res., Stat. Comput. Sci. (JORBEL), **33**(1–2), 79–102 (1992)

13. Brown, C.E., Sumichrast, R.T.: Impact of the replacement heuristic in a grouping genetic algorithm. Comput. & OR **30**(11), 1575–1593 (2003)

14. Halkidi, M., Batistakis, Y., Vazirgiannis, M.: On clustering validation techniques. J. Intell. Inf. Syst. **17**, 107–145 (2001)

15. Rousseeuw, P.: Silhouettes: a graphical aid to the interpretation and validation of cluster analysis. J. Comput. Appl. Math. **20**(1), 53–65 (1987)

16. Davies, D.L., Bouldin, D.W.: A cluster separation measure. IEEE Trans. Pattern Anal. Mach. Intell. **1**(2), 224–227 (1979)

17. Rand, W.: Objective criteria for the evaluation of clustering methods. J. Am. Stat. Assoc. **66**(336), 846–850 (1971)

18. Jain, A.K., Dubes, R.C.: Algorithms for Clustering Data. Prentice-Hall Inc., Upper Saddle River (1988)

19. Chang, D.X., Zhang, X.D., Zheng, C.W.: A genetic algorithm with gene rearrangement for k-means clustering. Pattern Recogn. **42**(7), 1210–1222 (2009)

20. MacQueen, J., et al.: Some methods for classification and analysis of multivariate observations. In: Proceedings of the Fifth Berkeley Symposium on Mathematical Statistics and Probability. Number 14 in 1, California, USA, pp. 281–297 (1967)

21. Dempster, A.P., Laird, N.M., Rubin, D.B.: Maximum likelihood from incomplete data via the EM algorithm. J. R. Stat. Soc., Ser. B **39**(1), 1–38 (1977)

22. Voorhees, E.M.: The effectiveness and efficiency of agglomerative hierarchical clustering in document retrieval. Ph.D. thesis (1985)

23. Kohonen, T.: The self-organizing map. Proc. IEEE **78**(9), 1464–1480 (1990)

24. Ankerst, M., Breunig, M.M., Peter Kriegel, H., Sander, J.: OPTICS: ordering points to identify the clustering structure. In: Delis, A., Faloutsos, C., Ghandeharizadeh, S. (eds.) Proceedings of the 1999 ACM SIGMOD International Conference on Management of Data, SIGMOD 1999, pp. 49–60. ACM Press (1999)

25. Madeira, S.C., Oliveira, A.L.: Biclustering algorithms for biological data analysis: a survey. IEEE/ACM Trans. Comput. Biol. Bioinform. **1**, 24–45 (2004)

26. Hong, Y., Kwong, S., Chang, Y., Ren, Q.: Unsupervised feature selection using clustering ensembles and population based incremental learning algorithm. Pattern Recogn. **41**(9), 2742–2756 (2008)

27. Özcan, E., Misir, M., Ochoa, G., Burke, E.K.: A reinforcement learning-great-deluge hyper-heuristic for examination timetabling. Int. J. Appl. Metaheuristic Comput. **1**(1), 39–59 (2010)

28. Burke, E.K., Bykov, Y.: The late acceptance hill-climbing heuristic. Eur. J. Oper. Res. **258**(1), 70–78 (2017)

29. Dueck, G.: New optimization heuristics: the great deluge algorithm and the record-to-record travel. J. Comput. Phys. **104**, 86–92 (1993)

30. Misir, M., Verbeeck, K., Causmaecker, P.D., Berghe, G.V.: A new hyper-heuristic as a general problem solver: an implementation in hyflex. J. Sched. **16**(3), 291–311 (2013)

31. Burke, E., Curtois, T., Hyde, M., Kendall, G., Ochoa, G., Petrovic, S., Vazquez-Rodriguez, J.: Hyflex: a flexible framework for the design and analysis of hyper-heuristics. In: Proceedings of the Multidisciplinary International Scheduling Conference (MISTA09), pp. 790–797 (2009)

32. Bache, K., Lichman, M.: UCI machine learning repository. School of Information and Computer Science, University of California, Irvine (2013)

Reference Point Adaption Method for Genetic Programming Hyper-Heuristic in Many-Objective Job Shop Scheduling

Atiya Masood[(✉)], Gang Chen[(✉)], Yi Mei[(✉)], and Mengjie Zhang[(✉)]

Victoria University of Wellington, Wellington, New Zealand
{masoodatiy,aaron.chen,yi.mei,mengjie.zhang}@ecs.vuw.ac.nz

Abstract. Job Shop Scheduling (JSS) is considered to be one of the most significant combinatorial optimization problems in practice. It is widely evidenced in the literature that JSS usually contains many (four or more) potentially conflicting objectives. One of the promising and successful approaches to solve the JSS problem is Genetic Programming Hyper-Heuristic (GP-HH). This approach automatically evolves dispatching rules for solving JSS problems. This paper aims to evolve a set of effective dispatching rules for many-objective JSS with genetic programming and NSGA-III. NSGA-III originally defines uniformly distributed reference points in the objective space. Thus, there will be few reference points with no Pareto optimal solutions associated with them; especially, in the cases with discrete and non-uniform Pareto front, resulting in many useless reference points during evolution. In other words, these useless reference points adversely affect the performance of NSGA-III and genetic programming. To address the above issue, in this paper a new reference point adaptation mechanism is proposed based on the distribution of the candidate solutions. We evaluated the performance of the proposed mechanism on many-objective benchmark JSS instances. Our results clearly show that the proposed strategy is promising in adapting reference points and outperforms the existing state-of-the-art algorithms for many-objective JSS.

Keywords: Job Shop Scheduling · Many-objective optimization
Genetic programming · Reference points

1 Introduction

Job Shop Scheduling (JSS) [11] is a classical combinatorial optimization problem that has received a lot of attention owing to its wide applicability in the real world. The JSS problem deals with the assignment of tasks or jobs to different resources or machines. The quality of schedule depends on the objective(s) of the JSS problem, e.g., the completion time or makespan.

It is widely noticeable in the literature [6] that JSS is a many-objective (i.e. more than three objectives) optimization problem. In [6] it was shown that

© Springer International Publishing AG, part of Springer Nature 2018
A. Liefooghe and M. López-Ibáñez (Eds.): EvoCOP 2018, LNCS 10782, pp. 116–131, 2018.
https://doi.org/10.1007/978-3-319-77449-7_8

makespan, mean flow time, maximum tardiness, mean tardiness and proportion of tardy jobs are potentially conflicting objectives.

JSS problem is widely known to be NP-hard [1] and dealing with many-objective increases its complexity. In practice, it is difficult to obtain an optimal schedule for large instances within a reasonable amount of time by using exact optimization methods. On the other hand, heuristic methods are fast and appropriate for practical scenarios. Conceptually, dispatching rules can be perceived as priority functions, used to assign priority to each and every waiting job.

Dispatching rules in JSS have shown promising performance [9]; however, designing dispatching rules is a complex and challenging research problem. Furthermore, dispatching rules often have inconsistent behaviors from one scenario to another. Genetic Programming Hyper-Heuristic (GP-HH) is a widely adopted technique in the literature that designs dispatching rules automatically under many conflicting objectives.

Although JSS has been proven to be many-objective optimization problem, only a few studies have focused on this issue. A recent work [6] proposed a new evolutionary many-objective optimization algorithm for JSS problems called GP-NSGA-III that seamlessly combines GP-HH with NSGA-III [3]. The proposed hybridized algorithm was compared to other approaches, GP-NSGA-II and GP-SPEA2 that combines GP-HH with multi-objective optimization algorithms including NSGA-II [2] and SPEA2 [14]. GP-NSGA-III significantly achieved better performance on 4-objective as well as 5-objective JSS problems compared to the other two approaches.

The adoption of uniformly distributed reference points in the GP-NSGA-III remains a challenge to the irregular and non-uniform Pareto front evolved in JSS problem. This is because many of these points are never associated with any dispatching rule on the evolved Pareto front resulting in many useless-reference points during evaluation. Evidently, useless references points negatively affect the performance of algorithms. This issue is also highlighted in [5,7] while applying NSGA-III to many-objective optimization problems, whose Pareto fronts are not uniform.

Another obvious drawback of GP-NSGA-III is the demote solution diversity, caused by a large number of useless reference points. In other words, GP-NSGA-III combines offspring and parents to form a set of solution of size 2N. After non-dominated sorting; a fixed set of N uniformly distributed reference points are used for the selection of the new population. In the scenario of irregular Pareto-front, each useful reference point is associated with a cluster of solutions that are existing in the closest proximity of that reference point. Thus, during Niching [3] selection of that useful points with heap of solutions may not help in exploration and directly reduce the solution diversity for future generation. Consequently, we need to find the better association between reference points and the evolved Pareto-front. This will help to enhance solution diversity as well as reduce the useless reference points.

In order to consider the useless reference point in GP-NSGA-III, we proposed the new reference point adaption mechanism. In this strategy, the whole objective space is decomposed into a small sub-simplex and reference points are generated according to the density of the solutions in that location. In other words, distribution of reference points follow the distribution of Pareto-front that establish better link between reference points and solutions in GP-NSGA-III.

The goal of this study to reduce the number of useless points by adaptively matching reference points with the evolved Pareto-front so as to promote the solution diversity and performance of GP-NSGA-III.

In the remainder of this paper, we first elaborate the research background including the JSS problem description and related works in Sect. 2. In Sect. 3, we propose an adaptive NSGA-III algorithm in detail. Section 4 covers the experimental design and parameter setting. Experimental results comparing the performance of the proposed algorithm with other three existing many-objectives JSS algorithms on training and test instances are presented in Sect. 5. Finally, Sect. 6 concludes this paper and highlights possible future research.

2 Background

In this section, the problem description will be presented first and then we will discuss some related works.

2.1 Problem Description of JSS

In a JSS problem, the shop includes a set of N jobs and M machines that need to be scheduled. In particular, each job j_i, $1 \leq i \leq N$ has a sequence of m operations and each operation u_i^k has its own pre-determined route through a sequence of machines $m_i^k, 1 \leq i \leq M$ with the fixed processing time $p_i^k > 0$. Any solution to such a JSS problem has to comply with four important constraints, as described below.

1. Operations are non-preemptive. This means that once an operation starts on a specific machine then the machine cannot be interrupted by any other operations.
2. The due date D_i for each job j_i must be set in advance and fixed during the execution of j_i.
3. Each M machine can process at most one operation from a queue at any given time.
4. Operations must follow precedence constraints, which means succeeding operations cannot be executed until all its previous operations have been completed.

In this study our goal is to evolve a set of non-dominating dispatching rules on the Pareto front. The quality of the schedule that was created by any specific rules will then be evaluated with respect to a number of objectives. In particular, following [6,7], we focus on minimizing four objectives, i.e. *mean flowtime (MF), maximal weighted tardiness (MaxWT), maximal flowtime (MaxF) and mean weighted tardiness (MWT)* in a many-objective JSS problem.

2.2 Related Work

Over the years, evolutionary multi-objective algorithms *(EMO)* are used for handling multiple-objective problems in a job shop [1]. In general, there are two alternative approaches used for solving multi-objective JSS, i.e. the aggregation method and the Pareto dominance method [10]. Among these approaches, the Pareto dominance concept has been successfully used for developing multi-objective JSS algorithms [6,10].

Eventhough EMO have been successfully used in a job shop, little effort [6,10] has been made in developing effective EC algorithm for many-objective JSS. To the best of our knowledge, GP-NSGA-III is the first algorithms designed for many-objective JSS and tried to address the issue of scalability and the so-called curse of dimensionality in many-objective JSS. GP-NSGA-III uses uniformly distributed reference points but does not work well with discrete optimization problems, such as the JSS problem [6]. This issue has opened a new research direction and has attracted substantial research attention.

Jain and Deb [4,5] recently proposed an interesting mechanisms for relocating reference points adaptively in *NSGAIII*. The proposed algorithm is called A-NSGA-III. Their strategy is able to adaptively include and exclude reference points which depend on the crowding region of current non-dominated front.

In the PSO-based relocation approach, GP-A-NSGA-III [7] relocates reference points toward the multiple global best particles which are some of the reference points with the most individuals associated with them.

These two algorithms, A-NSGA-III and GP-A-NSGAIII(PSO), mainly focus on reducing the useless reference points and improve the association between reference points and Pareto-optimal solutions. A-NSGA-III dynamically changes the original size of the reference points whereas GP-A-NSGAIII(PSO) does not change the total number of reference points during evolution. These two algorithms work very well on a number of optimization problems, but their reference point adjustment strategies still have limitations.

One of the limitations in ANSGA-III is also highlighted in [4]; A-NSGAIII does not allow any extra reference points around the corner points of hyperplane. If many solutions are associated with these boundary points then ultimately many useless points still exist in the algorithm and affect the performance of the algorithm. Moreover few high quality dispatching rules that are associated with these points may not be selected. These rules will be ignored in future generations because they can be far from the corresponding corner points. Due to the above reasons the algorithm may not improve the solution diversity in the algorithm.

In the case of *GP-A-NSGAIII (PSO)*, useless reference points in a swarm that are attracted towards the global best particles (particles those are associated with highest numbers of solutions). Each particle updates its position by using an updated velocity and moving to the proximity of the global best. Placement of the reference points to their new position without using a proper distributions strategy may lead to overlapping or clustering. Therefore many reference points still do not have a chance to be associated with any solutions and become useless points. Existence of useless points in the algorithm ultimately affects the performance of the algorithm.

In conclusion, existing algorithms still have an issue of useless reference points that can affect the performance of the algorithms. Our proposed approach is a density based approach that generates the reference points according to the distribution of the solutions on a simplex during evolution. Because the reference points follow the distribution of the solutions that allow better association and reduce the existence of the useless reference points. The proposed algorithm also alleviates the issue of corner points because we generate these reference points so that they must stay within a lower and upper bound of each sub-simplex carefully. Moreover, proposed algorithm does not add extra reference points during evolution and it is easy to implement in a high-dimensional objective space. It is particularly suitable for many-objective optimization problems, which have irregular Pareto front such as JSS.

3   Adaptive Reference Points for Many-Objective JSS

The basic framework of the proposed algorithm, GP-NSGA-III with Density Model based Reference Point Adaptation; (GP-NSGA-III-DRA) is described in Algorithm 1. The frame-work of the proposed algorithm is similar to the GP-NSGA-III. The proposed algorithm generates the reference points according to the distribution of solutions.

GP-NSGA-III-DRA starts with initial population P_0 and fitness evaluation of GP individuals that is described in Algorithm 2. Next, the predefined simplex locations $w_i \epsilon W$ are introduced, which partition the whole objective space in a number of independent sub-simplex developed by *Das and Dennis* systematic approach [3].

Each w_i has associated solutions that is calculated by perpendicular distance from each solution $\overline{s_i}$ in the normalized objective space. The solution $\overline{s_i}$ is closest to a location w_i is consider to be associated with w_i, detail is mentioned in Algorithm 3. Quantity of associated solutions in w_i represents the density of solutions of that location. Density of w_i helps to estimate the reference points of each location. Any probabilistic of selection mechanism can be used for estimating the reference points but in this algorithm *roulette wheel selection* is used particularly that can determine the number of required reference points in each sub-simplex; generation of reference points is described in Algorithm 4.

After knowing the demand of reference points in each sub-simplex; the reference points is generated on that specific location. Detail of reference points generation method is mentioned in Algorithm 4. In the following sub sections, each component the of proposed algorithm is described in detail.

3.1   Fitness Evaluation

Lines 1 and 6 of Algorithm 1 evaluate each *GP* individual by applying to a set of JSS training instances I_{train} as a dispatching rule. Then, the normalized

---

Algorithm 1. The framework of GP-NSGA-III-DRA.

Input : A training set I_{train}

Output: A set of non-dominated rules P^*

1 Initialize and evaluate the population P_0 of rules by the ramped-half-and-half method;

2 Generate the W that partition the Objective Space into sub-simplex locations;

3 Set $g \leftarrow 0$;

4 **while** $g < g_{max}$ **do**

5   Generate the offspring population Q_g using the crossover, mutation and reproduction of GP;

6   **foreach** $Q \in Q_g$ **do** Evaluate rule Q;

7     $R_g \leftarrow P_g \cup Q_g$;

8     $(F_1, F_2 \dots) = $ Non-dominated-sort(R_g);

9   *Repeat*;

10     $S_g = S_g \cup F_i$;

11     $i = i + 1$;

12     $until |S_g| \geq N$;

13   **if** $|S_g| = N$ **then**

14         $P_{(g+1)} = S_g$;

15   **end**

16   **if** $|S_g| > N$ **then**

17         $P_{(g+1)} = \cup_{j=1}^{l-1} F_j$;

18   **end**

19     $K = |F_l| : K = N - |P_{t+1}|$;

20   Normalize Objectives of members in S_g;

21     $\overline{S_g} = ObjectiveNormalization(S_g)$;

22   **foreach** $w \epsilon W$ **do**

23   identify member of $\overline{S_g}$ close to w;

24         $E(w) = Associatew(\overline{S_g}, W)$;

25   **end**

26     $Z_g^* = Generate(E(w), \overline{S_g}, W)$;

27   Construct the new population P_{g+1} by the $NSGA - III$ association and Niching;

28     $g \leftarrow g + 1$;

29 **end**

30 **return** *The non-dominated individuals* $P^* \subseteq P_{g_{max}}$;

---

objective values of the resultant schedules are set to its fitness values. The pseudo code of the fitness evaluation is given in Algorithm 2.

3.2   Reference Point Generation

The reference point generation scheme $Generate(Z_g, W)$ is described in Algorithm 4. In Algorithm 3, the density of the solution in each sub simplex

---

Algorithm 2. The fitness evaluation.

    Input : A training set I_{train} and an individual (rule) P
    Output: The fitness $f(P)$ of the rule P
1 **foreach** I *in* I_{train} **do**
2   Construct a schedule $\Delta(P, I)$ by applying the rule P to the JSS instance I;
3   Calculate the objective values $f(\Delta(P, I))$;
4 **end**
5 $f(P) \leftarrow \frac{1}{|I_{train}|} \sum_{I \in I_{train}} f(\Delta(P, I))$; ;
6 **return** $f(P)$;

Algorithm 3. *Associatew*$(\overline{S_g}, W)$

    Input : $\overline{S_g}, W$
    Output: $E(w)$ (individuals in the *wth* sub-simplex)
1 **foreach** $w \epsilon W$ **do**
2     $E(w) = \phi$;
3 **end**
4 **foreach** $s \epsilon \overline{S}_g$ **do**
5   **foreach** $w \epsilon W$ **do**
6   compute $d^{\perp}(s, w)$;
7   **end**
8   Assign $\hat{w} = argmin_{w \epsilon W} d^{\perp}(s, w)$;
9   Save s in $E(\hat{w})$;
10 **end**
11 **return** $E(w)$;

---

location $w_i \epsilon W$ is identified by minimum perpendicular distance. This density of the solution helps to find out the number of the required reference points in a specific location by using roulette wheel selection.

For generating the reference points the algorithm uses the average of the candidate solutions in the sub-simplex location w_i. Then the perpendicular distance between the center and existing solutions is calculated. The solution which is far from the centroid is selected first and a reference point is generated around the solution by using the mid point of center and the selected solution. Then, the next average is calculated from the remaining solutions that are still in the race of acquiring reference points. This is an iterative process that run until the desired number of reference points have been generated on the simplex.

It should be mentioned that, if one reference point is required in a sub-simplex then it should be the centroid. Moreover, the area of each sub-simplex is small and randomly distributed solutions are very close to each other. Therefore, the generated reference points from the center will almost cover the closest proximity of each solution and help to improve the association between solutions and reference points.

Algorithm 4. $Generate(E(w), \overline{S}_g, W)$

Input : $E(w), W, \overline{S}_g$

Output: Z_g^*

1 **foreach** $\hat{w} \epsilon W$ **do**
2   set quantity=0;
3   **foreach** $s \epsilon E(\hat{w})$ **do**
4   **if** s *associate with* $\hat{w}$ **then**
5   quantity=quantity+1;
6   **end**
7   **end**
8   Assign $D(\hat{w}) = quantity$;
9 **end**
10 **foreach** $D(\hat{w})$ **do**
11   Assign $P(\hat{w}) = \parallel D(\hat{w}) \parallel \div \parallel D \parallel$;
12   Assign $P(\hat{w}) = \parallel D(\hat{w}) \parallel$*length of reference points: or find out the required reference points of each sub simplex by roulette wheel selection;
13 **end**
14 **foreach** $\hat{w} \epsilon W$ **do**
15   set mean=0;
16   **foreach** $s \epsilon E(\hat{w})$ **do**
17   mean= $\parallel s \parallel$ + mean
18   **end**
19   mean= mean $\div \parallel S \parallel \epsilon \hat{w}$;
20   Assign Mean$(\hat{w})$= mean;
21 **end**
22 **foreach** $\hat{w} \epsilon W$ **do**
23   Assign $p = P(\hat{w})$;
24   set i=0;
25   **while** $i \leq p$ **do**
26   **foreach** $s \epsilon E(\hat{w})$ **do**
27   compute $d^{\perp}(s, Mean(\hat{w}))$;
28   **end**
29   Assign $\hat{s} = s : argmax_{d^{\perp}(s, Mean(\hat{w}))}$;
30   Assign $Z_g^* = (\hat{s} + Mean(\hat{w})) \div 2$;
31   Remove $\hat{s}$ from $E(\hat{w})$;
32   i=i+1;
33   **end**
34 **end**
35 **return** Z_g^*;

---

4   Experiment Design

In this section, we explain our design of experiments for the JSS problem and provide the configuration of the GP system, performance measure and the data set in detail.

4.1   Parameter Settings

Dispatching rules in our experiment adopt the tree based GP representation that is one of a common representation in literature [6]. These trees are constructed from the list of functions and terminals, as summarized in Tables 1 and 2. For all

Table 1. Functional sets for GP for JSS.

Function set	Meaning
$+$	Addition
$-$	Subtraction
$*$	Multiplication
$/$	Protected division operator
Max	Maximum
Min	Minimum
$if\text{-}then\text{-}else$	Conditional

Table 2. Terminal set of GP for JSS.

Attribute	Notation
Processing time of the operation	PT
Inverse processing time of the operation	IPT
Processing time of the next operation	NOPT
Ready time of the operation	ORT
Ready time of the next machine	NMRT
Work remaining	WKR
Number of operation remaining	NOR
Work in the next queue	WINQ
Number of operations in the next queue	NOINQ
Flow due date	FDD
Due date	DD
Weight	W
Number of operations in the queue	NOIQ
Work in the queue	WIQ
Ready time of the machine	MRT

compared algorithms, the initial GP population is created by the ramp-half-and-half method. The population size is set to 1024 to ensure that there is an enough diversity in the evolved solutions. The crossover, mutation and reproduction rates are set to 85%, 10% and 5% respectively. The maximal depth is set to 8. In each generation, individuals are selected using tournament selection with the size 7. The maximal number of generations is set to 51.

In our experiment, we compared our proposed algorithm with GP-ANSGA-III, GP-NSGA-III and GP-A-NSGA-III(PSO). Particularly, GP-NSGA-III is a recently developed algorithm specially designed for many-objective JSS problems but without adaptive reference points. GP-ANSGA-III is a well-known general purpose algorithm with adaptive reference points and GP-A-NSGA-III(PSO) is another adaptively relocate reference points by using the dynamics of PSO.

4.2   Data Set

To verify the effectiveness of our algorithm, we first apply on static JSS that is simpler than dynamic JSS [10]. Therefore, in this study we try to find an optimal solution in a static JSS problem. Our experiment requires static data set for measuring the performance of evolved heuristics. Hence, we selected Taillard (TA) static JSS benchmark instances [12] as a test bed. In a nutshell, it consists of 80 instances (ID from 1 to 80), divided into 8 subsets that cover JSS problems with a varied number of jobs (*15 to 100*) and number of machines (*5 to 20*).

In the experiments, each subset was further divided into training and test sets, each consisting of 40 instances. Since instances were static, release time of all the jobs was safely set to zero. The due date $dd(j_i)$ of each job j_i was calculated by *total workload strategy*. That is,

$$dd(j_i) = \lambda \times \sum_{k=1}^{m} p_i^k, \tag{1}$$

where λ is the due date factor, which is set to 1.3.

4.3   Performance Measures

In order to assess the performance of our proposed algorithm and compare with all algorithms in our experiments, we adopt two common measures, i.e. *Inverted Generational Distance* (IGD) [13] and *Hyper-Volume* (HV) [15]. These indicators describe how well the approximation of the true Pareto front is converged and diversified. Theoretically, a set of non-dominated solutions with better performance should require a smaller IGD value and larger HV value.

5   Results and Discussions

This section compares the obtained Pareto fronts from GP-NSGA-III, GP-A-NSGA-III(PSO based), GP-NSGA-III-DRA, GP-ANSGA-III on both training and test instances by using IGD [13] and HV [15]. Parallel coordinate plot shows the distribution of population and reference points.

5.1 Overall Result

During the GP search process, a rule was evaluated on the 40 training instances. The fitness function for each objective was found as the average normalized objective value [8]. For each algorithm, 30 independent GP runs obtained 30 final random sets of dispatching rules. Then, the rules were tested on the 40 test instances.

Tables 3 and 4 show the mean and standard deviation of the average HV and average IGD values of 30 Pareto fronts on the training instances. In addition, the Wilcoxon rank sum test with the significance level of 0.05 was conducted separately on both the HV and IGD of the rules obtained by the four algorithms. The result shows, if p-value is smaller than 0.05, then the best algorithm is considered significantly better than the other algorithms. The significantly better results were marked in bold.

Tables 3 and 4 reveal that GP-NSGA-III-DRA achieved significantly better performance in terms of HV as compared to the other algorithms. Also in terms of IGD, GP-NSGA-III-DRA clearly outperformed GP-ANSGAIII, GP-A-NSGA-III(PSO) and GP-NSGAII.

In regards to the test instances, Tables 5 and 6 exhibit the same pattern in terms of IGD. However, in terms of HV the proposed algorithm is better but not competitive with GP-A-NSGA-III and is significantly better than other two compared algorithms.

To have a better understanding of these algorithms' performance on the GP search process, we plotted (a) the HV of the non-dominated solutions obtained

Table 3. The mean and standard deviation over the average HV values of the 30 independent runs on Training instances of the compared algorithms in the 4-obj experiment. The significantly better results are shown in bold.

HV				
Statistic	GP-NSGA-III-DRA	GP-A-NSGA-III(PSO)	GP-ANSGA-III	GP-NSGA-III
Mean&(STD)	**0.71699(0.01103)**	0.67953(0.01847)‡	0.68867(0.01140)‡	0.676387(0.01165)‡

‡ and † indicate GP-NSGA-III-DRA performs significantly better than and equivalently to the corresponding algorithm, respectively.

Table 4. The mean and standard deviation over the average IGD values of the 30 independent runs on Training instances of the compared algorithms in the 4-obj experiment. The significantly better results are shown in bold.

IGD				
Statistic	GP-NSGA-III-DRA	GP-A-NSGA-III(PSO)	GP-ANSGA-III	GP-NSGA-III
Mean&(STD)	**0.001257(0.00014)**	0.00135(0.00027)‡	0.00155(0.00028)‡	0.00133(0.00011)‡

‡ and † indicate GP-NSGA-III-DRA performs significantly better than and equivalently to the corresponding algorithm, respectively.

Table 5. The mean and standard deviation over the average HV values of the 30 independent runs on Testing instances of the compared algorithms in the 4-obj experiment. The significantly better results are shown in bold.

HV				
Statistic	GP-NSGA-III-DRA	GP-A-NSGA-III(PSO)	GP-ANSGA-III	GP-NSGA-III
Mean&(STD)	**0.538581(0.02380)**	0.484011(0.02299)‡	0.502493(0.01693)†	0.45446(0.02964)‡

‡ and † indicate GP-NSGA-III-DRA performs significantly better than and equivalently to the corresponding algorithm, respectively.

Table 6. The mean and standard deviation over the average IGD values of the 30 independent runs on Testing instances of the compared algorithms in the 4-obj experiment. The significantly better results are shown in bold.

IGD				
Statistic	GP-NSGA-III-DRA	GP-A-NSGA-III(PSO)	GP-ANSGA-III	GP-NSGA-III
Mean&(STD)	**0.001734(0.00032)**	0.00184(0.00031)‡	0.002521(0.00026)‡	0.002277(0.00036)‡

‡ and † indicate GP-NSGA-III-DRA performs significantly better than and equivalently to the corresponding algorithm, respectively.

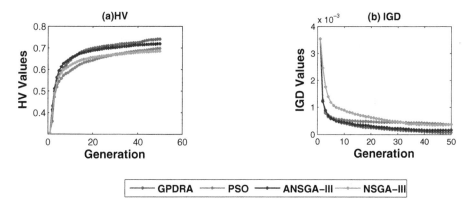

Fig. 1. HV and IGD values of the non-dominated solutions on the training set during the 30 independent GP runs.

so far and (b) the IGD of the non-dominated solutions obtained so far for each generation during the 30 independent runs of the four compared algorithms, as given in Fig. 1. These plots show that at early generations of evaluation, when solutions are not matured enough then our proposed algorithm had the same HV and IGD values than other compared algorithms. After a few generations, HV and IGD values of GP-NSGA-III-DRA are gradually improved as compared with GP-A-NSGA-III but significantly improved from GP-A-NSGA-III(PSO) and GP-NSGA-III. But in the last few generations, when the solutions moved toward the Pareto front, our proposed algorithm had significantly better HV and IGD as compared to the other algorithms. This suggests that generation by generation optimal solutions are improved in the case of GP-NSGA-III-DRA. This can also be seen in the parallel coordinate plot that shows the distribution

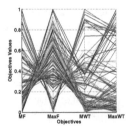

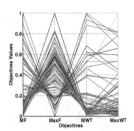

(a) Fitness values of the population in generation 10

(b) Fitness values of the population in generation 50

Fig. 2. Parallel coordinate plot for the fitness values of the population in generations 10 and 50 of NSGA-III.

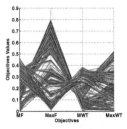

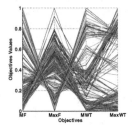

(a) Distribution of adaptive reference points in generation 10

(b) Fitness values of the population in generation 10

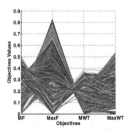

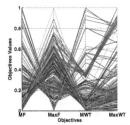

(c) Distribution of adaptive reference points in generation 50

(d) Fitness values of the population in generation 50

Fig. 3. Parallel coordinate plot for the distribution of the reference points and the fitness values of the population in generations 10 and 50 of GP-NSGA-III-DRA.

of the reference points and the fitness values of the population in generations on 10 and 50 of GP-A-NSGA-III(PSO), GP-ANSGA-III and GP-NSGA-III-DRA. GP-NSGA-III works on uniform distributed reference points, therefore we only show the distribution of the fitness value of GP-NSGA-III.

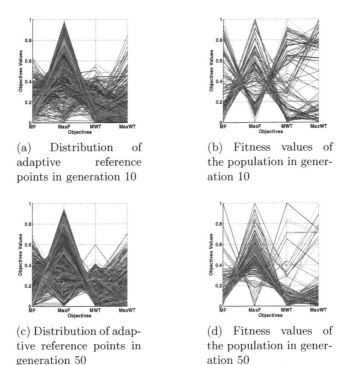

(a) Distribution of adaptive reference points in generation 10

(b) Fitness values of the population in generation 10

(c) Distribution of adaptive reference points in generation 50

(d) Fitness values of the population in generation 50

Fig. 4. Parallel coordinate plot for the distribution of the reference points and the fitness values of the population in generations 10 and 50 of GP-A-NSGA-III(PSO).

In Fig. 3 parallel coordinate plot shows that GP-NSGA-III-DRA has more diversified solutions as compared to the other three algorithms in Figs. 2, 4 and 5 because the distributions of the reference points and the fitness values of the population become very similar to each others. If we look at Fig. 4, we can observe that GP-A-NSGA-III(PSO) generated reference points on some locations where the simplex does not have any solution. These useless reference points may affect the performance of the algorithm. In the case of GP-ANSGA-III, Fig. 5 suggests that the distribution of the extra reference points is similar to the fitness value of population but these points are less than the desired number of reference points because the nature of the problem is many-objective [4]. Thus, uniform reference points are still a part of the reference set and affect the solution diversity of algorithm. In conclusion, GP-NSGA-III-DRA performance is better than the other algorithms because of better association and well diversified solutions on the Pareto-front that can easily observe in Fig. 1 and parallel coordinate plot.

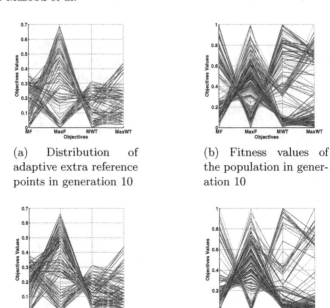

(a)   Distribution   of adaptive extra reference points in generation 10

(b)   Fitness values of the population in generation 10

(c)   Distribution   of adaptive extra reference points in generation 50

(d)   Fitness values of the population in generation 50

Fig. 5. Parallel coordinate plot for the distribution of the reference points and the fitness values of the population in generations 10 and 50 of GP-ANSGA-III.

6   Conclusion

In this paper, we have addressed a key research issue involved in using uniformly distributed reference points on the objective space for irregular Pareto front such as in many-objective JSS problem. An adaptive reference point strategy is proposed to enhance the performance of GP-NSGA-III. The proposed algorithm for generating the reference points according to the density of the solutions that provided more useful reference points. Better association between reference points and Pareto Optimal solutions will help to improve the usage of computational resources and promote the solution diversity. To examine the effectiveness and performance of our algorithm, experimental studies have been carried out by using the Taillard static job-shop benchmark set. Experimental studies suggest that GP-NSGA-III-DRA can perform significantly better than other compared algorithms in both training and testing instances and discover the Pareto-optimal solutions are more diversified than others compared algorithms. This finding leads us to believe that our algorithm GP-NSGA-III-DRA is more effective and efficient for many-objective JSS as compared to the existing algorithms. Rooms are still available for further improvement. For example, to achieve even a better match in between reference points and sampled solutions, we can utilize a Gaussian Process model to gradually and accurately learn the distribution of the Pareto front.

References

1. Błażewicz, J., Domschke, W., Pesch, E.: The job shop scheduling problem: conventional and new solution techniques. Eur. J. Oper. Res. **93**(1), 1–33 (1996)
2. Deb, K., Pratap, A., Agarwal, S., Meyarivan, T.: A fast and elitist multiobjective genetic algorithm: NSGA-II. IEEE Trans. Evol. Comput. **6**, 182–197 (2002)
3. Deb, K., Jain, H.: An evolutionary many-objective optimization algorithm using reference-point-based nondominated sorting approach, part I: solving problems with box constraints. IEEE Trans. Evol. Comput. **18**(4), 577–601 (2014)
4. Jain, H., Deb, K.: An improved adaptive approach for elitist nondominated sorting genetic algorithm for many-objective optimization. In: Purshouse, R.C., Fleming, P.J., Fonseca, C.M., Greco, S., Shaw, J. (eds.) EMO 2013. LNCS, vol. 7811, pp. 307–321. Springer, Heidelberg (2013). https://doi.org/10.1007/978-3-642-37140-0_25
5. Jain, H., Deb, K.: An evolutionary many-objective optimization algorithm using reference-point based nondominated sorting approach, part II: handling constraints and extending to an adaptive approach. IEEE Trans. Evol. Comput. **18**(4), 602–622 (2014)
6. Masood, A., Mei, Y., Chen, G., Zhang, M.: Many-objective genetic programming for job-shop scheduling. In: IEEE WCCI 2016 Conference Proceedings. IEEE (2016)
7. Masood, A., Mei, Y., Chen, G., Zhang, M.: A PSO-based reference point adaption method for genetic programming hyper-heuristic in many-objective job shop scheduling. In: Wagner, M., Li, X., Hendtlass, T. (eds.) ACALCI 2017. LNCS (LNAI), vol. 10142, pp. 326–338. Springer, Cham (2017). https://doi.org/10.1007/978-3-319-51691-2_28
8. Mei, Y., Zhang, M., Nyugen, S.: Feature selection in evolving job shop dispatching rules with genetic programming. In: GECCO. ACM (2016)
9. Nguyen, S., Zhang, M., Johnston, M.: A genetic programming based hyper-heuristic approach for combinatorial optimisation. In: Krasnogor, N., Lanzi, P.L. (eds.) GECCO, pp. 1299–1306. ACM (2011)
10. Nguyen, S., Zhang, M., Johnston, M., Tan, K.C.: Dynamic multi-objective job shop scheduling: a genetic programming approach. In: Uyar, A., Ozcan, E., Urquhart, N. (eds.) Automated Scheduling and Planning. SCI, vol. 505, pp. 251–282. Springer, Heidelberg (2013). https://doi.org/10.1007/978-3-642-39304-4_10
11. Pinedo, M.L.: Scheduling: Theory, Algorithms, and Systems. Springer Science & Business Media, Heidelberg (2012)
12. Taillard, E.: Benchmarks for basic scheduling problems. Eur. J. Oper. Res. **64**(2), 278–285 (1993)
13. Zhang, Q., Zhou, A., Zhao, S., Suganthan, P.N., Liu, W., Tiwari, S.: Multiobjective optimization test instances for the CEC 2009 special session and competition. Technical report. University of Essex, Colchester, UK and Nanyang technological University, Singapore, special session on performance assessment of multi-objective optimization algorithms, pp. 1–30 (2008)
14. Zitzler, E., Laumanns, M., Thiele, L.: SPEA2: improving the strength pareto evolutionary algorithm. In: EUROGEN 2001. Evolutionary Methods for Design, Optimization and Control with Applications to Industrial Problems, pp. 95–100 (2002)
15. Zitzler, E., Thiele, L., Laumanns, M., Fonseca, C.M., Da Fonseca, V.G.: Performance assessment of multiobjective optimizers: an analysis and review. IEEE Trans. Evol. Comput. **7**(2), 117–132 (2003)

MOEA/DEP: An Algebraic Decomposition-Based Evolutionary Algorithm for the Multiobjective Permutation Flowshop Scheduling Problem

Marco Baioletti[1], Alfredo Milani[1,2], and Valentino Santucci[1(✉)]

[1] Department of Mathematics and Computer Science, University of Perugia,
Perugia, Italy
{marco.baioletti,alfredo.milani,valentino.santucci}@unipg.it
[2] Department of Computer Science, Hong Kong Baptist University,
Kowloon Tong, Hong Kong

Abstract. Algebraic evolutionary algorithms are an emerging class of meta-heuristics for combinatorial optimization based on strong mathematical foundations. In this paper we introduce a decomposition-based algebraic evolutionary algorithm, namely MOEA/DEP, in order to deal with multiobjective permutation-based optimization problems. As a case of study, MOEA/DEP has been experimentally validated on a multiobjective permutation flowshop scheduling problem (MoPFSP). In particular, the makespan and total flowtime objectives have been investigated. Experiments have been held on a widely used benchmark suite, and the obtained results have been compared with respect to the state-of-the-art Pareto fronts for MoPFSP. The experimental results have been analyzed by means of two commonly used performance metrics for multiobjective optimization. The analysis clearly shows that MOEA/DEP reaches new state-of-the-art results for the considered benchmark.

Keywords: Algebraic evolutionary algorithms
Multiobjective optimization
Permutation Flowshop Scheduling Problem

1 Introduction and Related Work

Multiobjective optimization is receiving a growing level of interest in the evolutionary computation community (see for example [1–3]). Indeed, several real-world problems can be modeled using two or more objectives to be optimized simultaneously. Since the objectives may contrast each other, a population of somehow good (i.e., Pareto optimal) solutions is generally required by a decision maker. Hence, the innate ability of evolutionary algorithms to deal with populations of solutions explains their wide and successful application in multiobjective optimization.

© Springer International Publishing AG, part of Springer Nature 2018
A. Liefooghe and M. López-Ibáñez (Eds.): EvoCOP 2018, LNCS 10782, pp. 132–145, 2018.
https://doi.org/10.1007/978-3-319-77449-7_9

Recently, an algebraic framework has been successfully proposed as a general method to transform popular continuous meta-heuristics, such as Differential Evolution (DE) [4] and Particle Swarm Optimization (PSO) [5], to powerful algebraic evolutionary algorithms for combinatorial optimization. In particular, DEP (Differential Evolution for Permutations) has obtained state-of-the-art results on single-objective Permutation Flowshop Scheduling Problems (PFSPs) [4,6,7].

In this paper, basing on the widely known decomposition-based MOEA/D framework [8], we propose an algebraic evolutionary algorithm for multiobjective permutation-based optimization problems. Our algorithm, namely MOEA/DEP, is mainly based on the algebraic differential mutation operator [4,9]. As a case of study, we apply MOEA/DEP to a multiobjective PFSP (MoPFSP) [10] considering the makespan and total flowtime as objectives.

In MoPFSP, a set $J = \{1, \ldots, n\}$ of n jobs has to be scheduled on a set $M = \{1, \ldots, m\}$ of m machines. M is provided with a fixed order and each job has to visit, in the given order, all the machines. The processing time of job $j \in J$ on machine $i \in M$ is given by $p_{i,j}$. Every machine can process only one job at a time. Preemption and job-passing are not allowed. Hence, MoPFSP requires to find a permutation $\pi = \langle \pi(1), \ldots, \pi(n) \rangle$ of the n jobs that minimizes both the makespan (MS) and total flowtime (TFT) objective functions, defined, respectively, as

$$f_{MS} = c(m, \pi(n)), \tag{1}$$

$$f_{TFT} = \sum_{j=1}^{n} c(m, \pi(j)), \tag{2}$$

where, in both definitions, $c(i, \pi(j))$ is the completion time of the job $\pi(j)$ on the i-th machine and it is recursively computed as

$$c(i, \pi(j)) = p_{i,\pi(j)} + \max\{c(i, \pi(j-1)), c(i-1, \pi(j))\}, \tag{3}$$

when $i > 1$ and $j > 1$, while the terminal cases are:

$$\begin{aligned} c(1, \pi(1)) &= p_{1,\pi(1)}, \\ c(1, \pi(j)) &= p_{1,\pi(j)} + c(1, \pi(j-1)), \\ c(i, \pi(1)) &= p_{i,\pi(1)} + c(i-1, \pi(1)). \end{aligned} \tag{4}$$

A literature review for MoPFSP has been proposed in [10], where the performances of 23 algorithms have been compared. According to the authors, when MS and TFT are considered as objectives, the best performing algorithm is a multiobjective variant of the simulated annealing algorithm [11]. These results were later improved by RIPG [12] and TP+PLS [13]: both algorithms are mainly based on local search procedures hybridized with other heuristic mechanisms. However, to the best of our knowledge, current state-of-the-art results are those recently obtained in [14] by an estimation of distribution algorithm based on Mallows models, namely MEDA/D-MK. Therefore, we have experimentally compared our proposal with the reference Pareto fronts provided by the authors of [14].

The rest of the paper is organized as follows. Background concepts about multiobjective optimization and the popular MOEA/D framework are provided in Sect. 2. Section 3 briefly recalls the algebraic framework for combinatorial evolutionary algorithms. MOEA/DEP is introduced in Sect. 4, while the experimental analysis is provided in Sect. 5. Finally, conclusions are drawn in Sect. 6 where some future lines of research are also depicted.

2   Multiobjective Optimization and MOEA/D Framework

Without loss of generality, a Multiobjective Optimization Problem (MOP) can be represented as k objective functions $f_1, \ldots, f_k$ to minimize, each one defined on a *decision space* X, i.e., $f_i : X \to \mathbb{R}$ for all $i \in \{1, \ldots, k\}$.

Given two solutions $x, y \in X$, x *dominates* y, denoted by $x \prec y$, if and only if: $f_i(x) \leq f_i(y)$ for all $i \in \{1, \ldots, k\}$, and $f_i(x) < f_i(y)$ for at least one i. If the two solutions x and y are such that neither $x \prec y$ nor $y \prec x$, then they are *incomparable*.

A solution $x \in X$ is *Pareto optimal* if there exists no other solution $y \in X$ such that $y \prec x$. The *Pareto set* (*PS*) is the set of all the Pareto optimal solutions for a given MOP. It is easy to show that the solutions in *PS* are incomparable. The images of the *PS* solutions in the *objective space* are called *Pareto front* (*PF*), i.e., $PF = \{(f_1(x), \ldots, f_k(x)) | x \in PS\}$.

The goal of an algorithm for multiobjective optimization is to find a set of incomparable solutions that approximates as much as possible the Pareto front of the given MOP. Often, diversity in the objective space is also taken into account in order to provide a good coverage of the true Pareto front.

The population-based nature of Evolutionary Algorithms (EAs) makes them particularly suited in solving MOPs, hence a variety of multiobjective EAs (MOEAs) have been proposed [1,2,15]. The most popular classes of MOEAs are: (i) domination-based, (ii) indicator-based, and (iii) decomposition-based. In this paper we focus on the decomposition-based approach and, in particular, on the MOEA/D framework originally proposed in [8].

MOEA/D decomposes a MOP into several single objective subproblems such that every population individual optimizes its own subproblem. The population evolves and interacts by means of variation and replacement operators, while the approximated Pareto set is maintained as an external archive of incomparable solutions which are updated every time a new solution is generated.

MOEA/D requires a set of weight vectors and an aggregation function in order to generate N single objective subproblems and their neighborhoods.

The N weight vectors $\lambda_1, \ldots, \lambda_N$ are k-dimensional (where k is the number of objectives), their components are non-negative and sum-up to 1. With the aim of maintaining a certain degree of diversity, the weight vectors need to be evenly spread. In [8], the setting of N and $\lambda_1, \ldots, \lambda_N$ is controlled by a parameter H, thus that $\lambda_1, \ldots, \lambda_N$ are all the possible weight vectors whose components take a value from $\{\frac{0}{H}, \frac{1}{H}, \ldots, \frac{H}{H}\}$ and sum-up to 1. Hence, N and H are related by $N = \binom{H+k-1}{k-1}$.

The aggregation function has the form $g : X \times \mathbb{R}^k \to \mathbb{R}$. The two most popular choices for g are the weighted sum g^{ws} and the Tchebycheff function g_z^{tc} that, given a solution $x \in X$ and a weight vector $\lambda \in \mathbb{R}^k$, are respectively defined as

$$g^{ws}(x|\lambda) = \sum_{i=1}^{k} \lambda_i f_i(x), \tag{5}$$

$$g_z^{tc}(x|\lambda) = \max_{1 \le i \le k} \{\lambda_i | f_i(x) - z_i|\}, \tag{6}$$

where $z \in \mathbb{R}^k$ is the reference point in the objective space that it is ideally set such that $z_i = \min\{f_i(x)|x \in X\}$ for all $i \in \{1, \dots, k\}$. Practically, it is dynamically adjusted with the best objective values encountered so far in the evolution.

N population individuals, one for each weight vector, are deployed. The i-th individual is formed by: a candidate solution x_i, a weight vector λ_i, its fitness is given by $g(x_i|\lambda_i)$, and its neighborhood B_i is defined as $B_i = \{i_1, \dots, i_T\}$, where, for all $j \in B_i$, λ_j is among the T closest vectors to λ_i (by considering Euclidean distance).

The solutions $x_1, \dots, x_N$ are randomly initialized and the non-dominated ones enter the external archive EP. The populations $x_1, \dots, x_N$ and EP are iteratively evolved by means of genetic variation and replacement operators. For each individual i, a new solution x_i' is generated by using genetic operators, such as crossover and mutation, designed for the representation at hand. Usually, x_i' is obtained by recombining one or more solutions from $\{x_j | j \in \{i\} \cup B_i\}$. Then, x_i' replaces x_i if fitter in the i-th subproblem, i.e., when $g(x_i'|\lambda_i) < g(x_i|\lambda_i)$. Similarly, neighbors replacement is performed by comparing x_i' with its neighbors, i.e., for each index $j \in B_i$, x_i' replaces x_j if $g(x_i'|\lambda_j) < g(x_j|\lambda_j)$. Moreover, x_i' is also used to update the solutions in EP. At the end of the evolution, the approximated Pareto set is given by EP.

3 Algebraic Differential Evolution for Permutations

As described in [4], the design of the Algebraic Differential Evolution (ADE) resembles that of the classical DE. The population $\{x_1, \dots, x_N\}$ of N candidate solutions is iteratively evolved by means of the three operators of differential mutation, crossover and selection. Differently from numerical DE, ADE addresses combinatorial optimization problems whose search space is representable by finitely generated groups. Since crossover and selection schemes for combinatorial spaces are widely available in literature, the main focus is on the Differential Mutation (DM) operator. DM is widely recognized as the key component of DE [16] and, in its most common variant, generates a mutant v according to

$$v \leftarrow x_{r_0} \oplus F \odot (x_{r_1} \ominus x_{r_2}) \tag{7}$$

where $x_{r_0}, x_{r_1}, x_{r_2}$ are three randomly selected population individuals, while $F \in [0, 1]$ is the DE scale factor parameter. In numerical DE, the operators $\oplus, \ominus, \odot$ are the usual vector operations of $\mathbb{R}^n$, while, in ADE, their definitions are formally derived using the underlying algebraic structure of the search space.

The triple $(X, \circ, G)$ is a finitely generated group representing a combinatorial search space if:

- X is the discrete set of solutions;
- $\circ$ is a binary operation on X satisfying the group properties, i.e., closure, associativity, identity (e), and invertibility (x^{-1}); and
- $G \subseteq X$ is a finite generating set of the group, i.e., any $x \in X$ has a (not necessarily unique) minimal-length decomposition $\langle g_1, \ldots, g_l \rangle$, with $g_i \in G$ for all $i \in \{1, \ldots, l\}$, and whose evaluation is x, i.e., $x = g_1 \circ \cdots \circ g_l$.

For the sake of clarity, the length of a minimal decomposition of x is denoted with $|x|$.

Using $(X, \circ, G)$ we can provide the formal definitions of the operators $\oplus, \ominus, \odot$. Let $x, y \in X$ and $\langle g_1, \ldots, g_k, \ldots, g_{|x|} \rangle$ be a minimal decomposition of x, then

$$x \oplus y := x \circ y, \tag{8}$$

$$x \ominus y := y^{-1} \circ x, \tag{9}$$

$$F \odot x := g_1 \circ \cdots \circ g_k, \text{ with } k = \lceil F \cdot |x| \rceil \text{ and } F \in [0, 1]. \tag{10}$$

The algebraic structure on the search space naturally defines neighborhood relations among the solutions. Indeed, it induces a colored digraph whose vertices are the solutions in X and two generic solutions $x, y \in X$ are linked by an arc with color $g \in G$ if and only if $y = x \circ g$. Therefore, a simple one-step move can be directly encoded by a generator, while a composite move can be synthesized as the evaluation of a sequence of generators (a path on the graph). In analogy with $\mathbb{R}^n$, the elements of X can be dichotomously interpreted both as solutions (vertices on the graph) and as displacements between solutions (colored paths on the graph). As detailed in [4], this allows to provide a rational interpretation to the discrete DM of definition (7). The key idea is that the difference $x \ominus y$ is the evaluation of the generators on a shortest path from y to x.

Clearly, $\oplus$ and $\ominus$ do not depend on the generating set, thus they are uniquely defined. Conversely, $\odot$ relies on the chosen generating set and, also fixing it, a minimal decomposition is not unique in general. Therefore, $\odot$ is implemented as a stochastic operator, thus requiring a randomized decomposition algorithm for the finitely generated group at hand.

The algebraic Differential Evolution for Permutations (DEP) [4] is an implementation of ADE for the search space of permutations. Indeed, the permutations of the set $\{1, \ldots, n\}$, together with the usual permutation composition, form the so-called Symmetric group $\mathcal{S}(n)$. Since $\mathcal{S}(n)$ is finite, it is also finitely generated. Different choices for the generating set are possible. The most common and practical ones are:

- ASW, which is based on *adjacent swap moves* and it is formally defined as

$$ASW = \{\sigma_i : 1 \leq i < n\}, \tag{11}$$

where σ_i is the identity permutation with the items i and $i+1$ exchanged. Hence, given a generic $\pi \in \mathcal{S}_n$, the composition $\pi \circ \sigma_i$ swaps the i-th and $(i+1)$-th items of π.

- *EXC*, which is based on *exchange moves* and it is formally defined as

$$EXC = \{\epsilon_{ij} : 1 \leq i < j \leq n\}, \tag{12}$$

where ϵ_{ij} is the identity permutation with the items i and j exchanged. Hence, given a generic $\pi \in \mathcal{S}_n$, the composition $\pi \circ \epsilon_{ij}$ swaps the i-th and j-th items of π.

- *INS*, which is based on *insertion moves* and it is formally defined as

$$INS = \{\iota_{ij} : 1 \leq i, j \leq n\}, \tag{13}$$

where ι_{ij} is the identity permutation where the item i is shifted to position j. Hence, given a generic $\pi \in \mathcal{S}_n$, the composition $\pi \circ \iota_{ij}$ shifts the i-th item in π to position j.

A minimal decomposition for a generic permutation $\pi \in \mathcal{S}_n$ can be obtained by ordering the items in π using a sorting algorithm that only performs moves from the chosen set, i.e., adjacent swaps for *ASW*, generic exchanges for *EXC*, and insertions for *INS*. The sequence of generators corresponding to the moves performed during the sorting process is annotated, then the sequence is reversed and each generator is replaced with its inverse [4].

Therefore, randomized decomposers for *ASW*, *EXC*, and *INS* have been implemented by means of generalized and randomized variants of, respectively, the bubble-sort, selection-sort, and insertion-sort algorithms [4,9]. They have been called *RandBS*, *RandSS*, *RandIS* and each one exploits a different algebraic property of permutations. Since (the ordered permutation) e is the only permutation with 0 inversions, *RandBS* iteratively decreases the number of inversions by swapping two suitable adjacent items. *RandSS* exploits the fact that e is the only permutation with n cycles in its cycle decomposition, thus it iteratively increases the number of cycles by exchanging two items belonging to the same cycle[1]. *RandIS* considers that e is the only permutation with a (unique) longest increasing subsequence (LIS) of length n, thus it iteratively extends a LIS by shifting a suitable item in a suitable position.

For further implementation details, proofs of correctness and complexity we refer the interested reader to [4,9].

4 MOEA/DEP

In this section we introduce MOEA/DEP: an algebraic evolutionary algorithm for permutation-based multiobjective optimization problems. MOEA/DEP

---

[1] In our previous paper [9] there was a typo. Actually, the time complexity of *RandSS* is $\Theta(n)$ and not $\Theta(n^2)$ as erroneously in [9]. A simple amortized analysis proves the claim.

implements the MOEA/D framework, described in Sect. 2, by relying on the algebraic operators for the permutations space described in Sect. 3.

Given a MOP represented by k objective functions such as $f_i : \mathcal{S}_n \to \mathbb{R}$ for all $i \in \{1, \ldots, k\}$, MOEA/DEP decomposes the MOP into N single objective subproblems and simultaneously minimizes them by evolving a population of N permutations $\pi_1, \ldots, \pi_N \in \mathcal{S}_n$ that are also used to build up the approximated Pareto set in the archive EP.

The algorithmic scheme of MOEA/DEP is depicted in Algorithm 1. Various initializations are performed at lines 1–4, then the population is evolved in the generational loop of lines 5–23. Every iteration can be divided in two parts: offsprings generation (lines 6–12), and population update (lines 13–22). When a given termination criterion is met, MOEA/DEP returns the approximated Pareto set EP.

Algorithm 1. MOEA/DEP scheme

1: Initialize N evenly distributed weight vectors $\lambda_1, \ldots, \lambda_N$
2: For all $i \in \{1, \ldots, N\}$, compute the B_i neighborhood of size T
3: Initialize N permutations (solutions) $\pi_1, \ldots, \pi_N$
4: Initialize EP with the non-dominated solutions in $\{\pi_1, \ldots, \pi_N\}$
5: **while** termination condition is not met **do**
6:     **for** each subproblem $i \in \{1, \ldots, N\}$ **do**
7:         Set $C_i \leftarrow \{1, \ldots, N\} \setminus \{i\}$ with probability q, or B_i otherwise
8:         Randomly select three distinct indices r_0, r_1, r_2 from C_i
9:         Set the "base permutation" $\rho_i \leftarrow \pi_{r_0}$ and perturbate it with probability pm
10:        Generate the mutant $\nu_i \leftarrow \rho_i \oplus F \odot (\pi_{r_1} \ominus \pi_{r_2})$ with the chosen $gset$
11:        Generate the offspring $\pi_i' \leftarrow CrossOver(\pi_i, \nu_i, CR)$
12:    **end for**
13:    **for** each subproblem $i \in \{1, \ldots, N\}$ **do**
14:        Evaluate $f_1(\pi_i'), \ldots, f_k(\pi_i')$ and update EP
15:        Update π_i:
16:            **if** $g(\pi_i'|\lambda_i) < g(\pi_i|\lambda_i)$, **then** $\pi_i \leftarrow \pi_i'$
17:        Update neighbors (no more than U):
18:            set $u \leftarrow 0$
19:            **while** $C_i \neq \emptyset$ and $u < U$ **do**
20:                $j \leftarrow$ select and remove a random index from C_i
21:                **if** $g(\pi_i'|\lambda_j) < g(\pi_j|\lambda_j)$, **then** $\pi_j \leftarrow \pi_i'$ and $u \leftarrow u + 1$
22:    **end for**
23: **end while**
24: **return** EP

Detailed descriptions of the initialization, offsprings generation, and population update are provided in, respectively, Sects. 4.1, 4.2, and 4.3, while a short description of the employed crossover operators is provided in Sect. 4.4.

4.1   Initialization

The initializations performed at lines 1–4 have been implemented following the original MOEA/D scheme [8] described in Sect. 2. Moreover, the initial individuals $\pi_1, \ldots, \pi_N$ are sampled uniformly random among all the permutations in $\mathcal{S}_n$, though, optionally, one of them can be initialized using a heuristic function for the problem at hand (e.g., the Liu-Reeves constructive heuristic for PFSP [17]).

4.2 Offsprings Generation

At lines 6–12, starting from the current individuals $\pi_1, \ldots, \pi_N$, a population of offsprings $\pi_1', \ldots, \pi_N'$ is generated by means of the differential mutation (DM) and crossover operators.

Two variants with respect to the original MOEA/D and DM schemes have been considered:

1. A (possibly) *extended neighborhood*: at lines 7–8, the three individuals undergoing DM are randomly selected from the whole population with probability q, or from the neighbors in B_i with probability $1 - q$.
2. A *pre-mutation* operation: at line 9, the so-called "base individual" of DM undergoes a preliminary small perturbation with probability pm. The perturbation is implemented as a single random insertion move, i.e., $\rho_i \leftarrow \rho_i \circ \iota$, where ι is randomly selected from INS.

Both the extended neighborhood and the pre-mutation allow to mitigate the loss of diversity in the population. Indeed, since $\pi \odot F(\pi \ominus \pi) = \pi$ for any $\pi \in \mathcal{S}_n$ and $F \in [0, 1]$, the DM operator cannot generate novel genotypes once the pool from which it selects its input individuals is converged to a super-genotype. Hence, the extended neighborhood mitigates this problem by also allowing the selection from the whole population. Moreover, if also the whole population is converged to a single-genotype, the pre-mutation allows anyway to make small moves.

Basing on the chosen generating set $gset \in \{ASW, EXC, INS\}$, at line 10, the mutant ν_i is generated by means of the algebraic DM operator described in Sect. 3.

Finally, at line 11, an offspring π_i' is obtained by recombining the mutant ν_i with the original individual π_i. The parameter CR may, or may not, be considered basing on the chosen crossover operator. See Sect. 4.4 for their definitions.

4.3 Population Update

At lines 13–22, both the main population $\pi_1, \ldots, \pi_N$ and the external archive EP are updated by considering the newly generated offsprings $\pi_1', \ldots, \pi_N'$.

EP is updated using a non-dominance criterion. A new solution π' enters EP if and only if there exists no $\rho \in EP$ such that $\rho \prec \pi'$. Once π' is added to EP, every $\rho \in EP$ such that $\pi' \prec \rho$ is removed from EP. Hence, at any moment of the evolution, EP contains the set of incomparable solutions that are not dominated by any of the solutions generated so far.

The main population individuals $\pi_1, \ldots, \pi_N$ are updated basing on their own single objective subproblem. Basically, the MOEA/D scheme described in Sect. 2 is applied with only few variants.

Since the k MOP objective functions may have different scales, as in [14,18], we have used a normalization approach for both the weighted sum and the Tchebycheff aggregation functions. Hence, the Eqs. (5) and (6) are replaced with

$$g_{z,w}^{ws}(x|\lambda) = \sum_{i=1}^{k} \lambda_i \frac{f_i(x) - \alpha z_i}{w_i - z_i}, \tag{14}$$

$$g_{z,w}^{tc}(x|\lambda) = \max_{1 \le i \le k} \left\{ \lambda_i \frac{f_i(x) - \alpha z_i}{w_i - z_i} \right\}, \tag{15}$$

where each z_i and w_i are, respectively, the minimum (best) and the maximum (worst) values observed so far for the i-th objective function, while an $\alpha < 1$ guarantees that αz_i is a proper lower bound. As in [14,18], we set $\alpha = 0.6$.

Therefore, basing on the chosen aggregation function $g \in \{g^{ws}, g^{tc}\}$, the population is updated similarly to the general framework described in Sect. 2. Note that the (possibly) extended neighborhood introduced in Sect. 4.2 influences also the neighbors replacement. Moreover, for the sake of population diversity, a *limited neighborhood update* has been introduced, i.e., every offspring is allowed to replace no more than U neighbors (that are randomly scanned).

4.4 Crossover for Permutations

In this work we have experimented two popular crossovers for permutation representations, namely, the two point crossover *TPII* adopted in [4] and the order based crossover *OBX* used in [9,19].

Given the parents $\pi, \nu \in \mathcal{S}(n)$, both *TPII* and *OBX* select a random subset of positions $P \subseteq \{1, \ldots, n\}$ and build the offspring $\pi' \in \mathcal{S}(n)$ by setting $\pi'(i) \leftarrow \pi(i)$ for any $i \in P$, and inserting the remaining items starting from the leftmost free place of π' and following the order of appearance in ν.

The difference between *TPII* and *OBX* is that *TPII* uses an interval of contiguous positions, while, in *OBX*, P can be any subset.

Furthermore, two additional variants of *TPII* and *OBX* have been introduced in order to consider the classical DE crossover parameter $CR \in [0, 1]$. The modified variants, $TPII^{CR}$ and OBX^{CR}, constrain the size of P to $|P| = \lceil CR \cdot n \rceil$.

Therefore, in line 11 of Algorithm 1, *CrossOver* is chosen from $\{TPII, OBX, TPII^{CR}, OBX^{CR}\}$.

5 Experiments

MOEA/DEP has been experimentally investigated on the multiobjective PFSP (MoPFSP) considering the two popular objectives MS and TFT defined, respectively, in Eqs. (1) and (2).

Experiments have been held on the widely known Taillard benchmark suite for PFSP [20] composed by 110 instances, i.e., 10 instances for each combination of $n = \{20, 50, 100, 200\}$ (number of jobs) and $m = \{5, 10, 20\}$ (number of machines), except the combination $n = 200, m = 5$.

The effectiveness of MOEA/DEP is compared with the recent state-of-the-art algorithm for MoPFSP, i.e., the estimation of distribution algorithm MEDA/D-MK introduced in [14]. In order to perform a fair comparison, the same performance metrics used in [14] have been considered, i.e., the hypervolume (HV) [21] and the coverage indicator (C-metric) [22].

The hypervolume is computed by means of a reference point $w \in \mathbb{R}^k$ in the objective space dominated by all the considered solutions. Hence, given a set of solutions A, $HV(A)$ measures the volume of the part of objective space dominated by A and bounded by w. The exact formulation of HV is provided in [21]. Moreover, as done in [14], the Pareto fronts have been normalized using the same mechanism of Eqs. (14) and (15), while w is set to $(1.01, 1.01)$.

The C-metric compares two solutions sets A and B by counting the relative number of solutions in B that are dominated by at least one solution in A, i.e., $C(A, B) = |\{b \in B | \exists a \in A : a \prec b\}| / |B|$. Note that C is not symmetric, thus $C(A, B)$ is usually compared with $C(B, A)$ in order to establish which is the best solutions set.

Also the same termination criterion of [14] has been adopted, i.e., every MOEA/DEP execution terminates after $n \times 1000$ generations have been performed.

In the following two subsections we describe the MOEA/DEP parameters' calibration and the results of the comparison with MEDA/D-MK.

5.1 Parameters Calibration

MOEA/DEP has different components and parameters that need to be appropriately tuned.

One of the initial individual is generated by means of the popular Liu-Reeves procedure $LR(n/m)$ [17] used in variety of PFSP algorithms (see for example [4,14]), while the other ones are randomly initialized. The differential evolution parameters F and CR are self-adapted during the evolution by means of the popular self-adaptive scheme jDE [23]. After some preliminary experiments we set $N = 100$, $T = 20$, $U = 2$, while the other parameters have been experimentally tuned by means of a full factorial analysis. The chosen ranges are:

- $g \in \{g^{ws}, g^{tc}\}$;
- $q \in \{0.25, 0.5, 0.75\}$;
- $pm \in \{0.7, 0.85, 1\}$;
- $gset \in \{ASW, EXC, INS\}$;
- $CrossOver \in \{TPII, OBX, TPII^{CR}, OBX^{CR}\}$.

Hence, a total of 216 MOEA/DEP settings have been executed on the first instance of every $n \times m$ problem configuration by performing 10 executions per setting. The total number of executions amounts to 23 760.

For every run, the hypervolume of the obtained Pareto set is computed. Then, on every instance $inst$, the hypervolumes of the same algorithm setting alg have been aggregated using the average relative percentage deviation measure

$$ARPD^{alg}_{inst} = \frac{1}{10} \sum_{i=1}^{10} \frac{\left(HV^{best}_{inst} - HV^{alg,i}_{inst}\right)}{HV^{best}_{inst}} \times 100, \tag{16}$$

where $HV^{alg,i}_{inst}$ is the hypervolume obtained by the setting alg on its i-th run on the instance $inst$, while HV^{best}_{inst} is the best hypervolume obtained by any setting in any run on the instance $inst$.

The setting with the best Quade average rank [24] computed on the ARPD values is chosen. Briefly, in any instance a ranking of the algorithms/settings is obtained basing on its ARPD value. The instances are weighted basing on the variability of its ARPDs. Finally, the ranks of an algorithm on the different instances are aggregated using a weighted average. For further details, see [24].

The winning MOEA/DEP configuration is

$$g = g^{ws}, q = 0.75, pm = 0.7, gset = INS, CrossOver = TPII. \tag{17}$$

5.2   Comparison with MEDA/D-MK

MEDA/D-MK has been recently proposed in [14] where it has been shown to be the state-of-the-art algorithm for the MoPFSP with the objectives MS and TFT.

In order to compare MOEA/DEP with MEDA/D-MK, the tuned MOEA/DEP setting (see Sect. 5.1) has been executed, 20 times per instance, on the 110 benchmark instances, while the Pareto fronts of MEDA/D-MK have been obtained from the website of their authors[2]. Moreover, as done in [14], given an instance, the 20 Pareto fronts obtained by MOEA/DEP have been aggregated into a single Pareto front by merging them and removing the dominated solutions.

The comparison between the Pareto fronts has been performed using both the hypervolume and the C-metric. The experimental results, averaged for any $n \times m$ problem configuration, are provided in Table 1, where A and B denote, respectively, MOEA/DEP and MEDA/D-MK.

Table 1 clearly shows that MOEA/DEP systematically outperforms MEDA/D-MK. The difference in terms of hypervolumes is particularly large on the instances with $n = 200$, though it is never below 0.2. Moreover, remarkable results are obtained also considering C-metric. In general, around 90% of the solutions in the Pareto sets of MEDA/D-MK are dominated by our results.

Finally, note that also the Pareto fronts obtained in a single run of MOEA/DEP are generally better than the reference ones, though with a slightly smaller difference with respect to the data provided in Table 1.

---

[2] https://github.com/murilozangari.

Table 1. Comparison between $A = $ MOEA/DEP and $B = $ MEDA/D-MK

Problem	HV_A	HV_B	$C(A, B)$	$C(B, A)$
20 × 5	**0.884**	0.612	**0.988**	0
20 × 10	**0.901**	0.640	**0.997**	0
20 × 20	**0.889**	0.648	**0.993**	0
50 × 5	**0.981**	0.730	**0.995**	0
50 × 10	**0.904**	0.623	**0.937**	0
50 × 20	**0.906**	0.635	**0.938**	0.003
100 × 5	**0.984**	0.597	**0.989**	0
100 × 10	**0.944**	0.521	**0.949**	0
100 × 20	**0.885**	0.622	**0.884**	0
200 × 10	**0.959**	0.497	**0.997**	0
200 × 20	**0.862**	0.547	**0.882**	0

6   Conclusion and Future Work

In this paper, basing on the recently proposed algebraic framework for combinatorial optimization, we have introduced MOEA/DEP: a decomposition-based algebraic evolutionary algorithm for multiobjective permutation-based problems.

The selection and replacement mechanisms of MOEA/DEP are based on the MOEA/D framework for multiobjective optimization. Moreover, in order to mitigate the diversity loss during the evolution, MOEA/DEP introduces some additional components and variants such as: a pre-mutation procedure, a (possibly) extended neighborhood for individuals selection, and a limited update of the neighboring individuals.

Experiments have been held on the multiobjective permutation flowshop scheduling problem (MoPFSP) considering the two popular criteria: makespan and total flowtime. A commonly used benchmark suite has been considered. The MOEA/DEP parameters have been properly calibrated, and the experimental results have been compared with respect to recent state-of-the-art results for MoPFSP [14]. The experimental analysis has been conducted using two popular performance metrics for multiobjective optimization. The experimental results clearly show that MOEA/DEP systematically outperforms the previous state-of-the-art algorithm.

Starting from the significant results obtained in this work, future lines of research will include: (i) the application of MOEA/DEP to other MoPFSP objectives and to other permutation-based multiobjective problems (e.g., see [25,26]), (ii) the use of the three generating sets altogether by means of a self-adaptive scheme.

References

1. Trivedi, A., Srinivasan, D., Sanyal, K., Ghosh, A.: A survey of multiobjective evolutionary algorithms based on decomposition. IEEE Trans. Evol. Comput. **21**(3), 440–462 (2017)
2. Zitzler, E., Thiele, L.: Multiobjective evolutionary algorithms: a comparative case study and the strength pareto approach. IEEE Trans. Evol. Comput. **3**(4), 257–271 (1999)
3. Zitzler, E., Deb, K., Thiele, L.: Comparison of multiobjective evolutionary algorithms: empirical results. Evol. Comput. **8**(2), 173–195 (2000)
4. Santucci, V., Baioletti, M., Milani, A.: Algebraic differential evolution algorithm for the permutation flowshop scheduling problem with total flowtime criterion. IEEE Trans. Evol. Comput. **20**(5), 682–694 (2016). https://doi.org/10.1109/TEVC.2015.2507785
5. Baioletti, M., Milani, A., Santucci, V.: Algebraic particle swarm optimization for the permutations search space. In: Proceedings of IEEE Congress on Evolutionary Computation CEC 2017, pp. 1587–1594 (2017). https://doi.org/10.1109/CEC.2017.7969492
6. Santucci, V., Baioletti, M., Milani, A.: Solving permutation flowshop scheduling problems with a discrete differential evolution algorithm. AI Commun. **29**(2), 269–286 (2016). https://doi.org/10.3233/AIC-150695
7. Santucci, V., Baioletti, M., Milani, A.: A differential evolution algorithm for the permutation flowshop scheduling problem with total flow time criterion. In: Bartz-Beielstein, T., Branke, J., Filipič, B., Smith, J. (eds.) PPSN 2014. LNCS, vol. 8672, pp. 161–170. Springer, Cham (2014). https://doi.org/10.1007/978-3-319-10762-2_16
8. Zhang, Q., Li, H.: MOEA/D: a multiobjective evolutionary algorithm based on decomposition. IEEE Trans. Evol. Comput. **11**(6), 712–731 (2007)
9. Baioletti, M., Milani, A., Santucci, V.: An extension of algebraic differential evolution for the linear ordering problem with cumulative costs. In: Handl, J., Hart, E., Lewis, P.R., López-Ibáñez, M., Ochoa, G., Paechter, B. (eds.) PPSN 2016. LNCS, vol. 9921, pp. 123–133. Springer, Cham (2016). https://doi.org/10.1007/978-3-319-45823-6_12
10. Minella, G., Ruiz, R., Ciavotta, M.: A review and evaluation of multiobjective algorithms for the flowshop scheduling problem. INFORMS J. Comput. **20**(3), 451–471 (2008)
11. Varadharajan, T., Rajendran, C.: A multi-objective simulated-annealing algorithm for scheduling in flowshops to minimize the makespan and total flowtime of jobs. Eur. J. Oper. Res. **167**(3), 772–795 (2005)
12. Minella, G., Ruiz, R., Ciavotta, M.: Restarted iterated Pareto Greedy algorithm for multi-objective flowshop scheduling problems. Comput. Oper. Res. **38**(11), 1521–1533 (2011)
13. Dubois-Lacoste, J., López-Ibáñez, M., Stützle, T.: A hybrid TP+PLS algorithm for bi-objective flow-shop scheduling problems. Comput. Oper. Res. **38**(8), 1219–1236 (2011)
14. Zangari, M., Mendiburu, A., Santana, R., Pozo, A.: Multiobjective decomposition-based mallows models estimation of distribution algorithm. A case of study for permutation flowshop scheduling problem. Inf. Sci. **397–398**(Supplement C), 137–154 (2017)

15. Deb, K., Pratap, A., Agarwal, S., Meyarivan, T.: A fast and elitist multiobjective genetic algorithm: NSGA-II. IEEE Trans. Evol. Comput. **6**(2), 182–197 (2002)
16. Storn, R., Price, K.: Differential evolution - a simple and efficient heuristic for global optimization over continuous spaces. J. Glob. Optim. **11**(4), 341–359 (1997)
17. Liu, J., Reeves, C.R.: Constructive and composite heuristic solutions to the$P//\Sigma C_i$ scheduling problem. Eur. J. Oper. Res. **132**(2), 439–452 (2001)
18. Chang, P.C., Chen, S.H., Zhang, Q., Lin, J.L.: MOEA/D for flowshop scheduling problems. In: Proceedings of IEEE Congress on Evolutionary Computation CEC 2008, pp. 1433–1438, June 2008
19. Baioletti, M., Milani, A., Santucci, V.: Linear ordering optimization with a combinatorial differential evolution. In: Proceedings of 2015 IEEE International Conference on Systems, Man, and Cybernetics, SMC 2015, pp. 2135–2140 (2015). https://doi.org/10.1109/SMC.2015.373
20. Taillard, E.: Benchmarks for basic scheduling problems. Eur. J. Oper. Res. **64**(2), 278–285 (1993)
21. Beume, N., Fonseca, C.M., López-Ibáñez, M., Paquete, L., Vahrenhold, J.: On the complexity of computing the hypervolume indicator. IEEE Trans. Evol. Comput. **13**(5), 1075–1082 (2009)
22. Schutze, O., Esquivel, X., Lara, A., Coello, C.A.C.: Using the averaged Hausdorff distance as a performance measure in evolutionary multiobjective optimization. IEEE Trans. Evol. Comput. **16**(4), 504–522 (2012)
23. Brest, J., Greiner, S., Boskovic, B., Mernik, M., Zumer, V.: Self-adapting control parameters in differential evolution a comparative study on numerical benchmark problems. IEEE Trans. Evol. Comput. **10**(6), 646–657 (2006)
24. Derrac, J., García, S., Molina, D., Herrera, F.: A practical tutorial onthe use of nonparametric statistical tests as a methodology for comparing evolutionary and swarm intelligence algorithms. Swarm Evol. Comput. **1**(1), 3–18 (2011)
25. Santucci, V., Baioletti, M., Milani, A.: An algebraic differential evolution for the linear ordering problem. In: Companion Material Proceedings of Genetic and Evolutionary Computation Conference, GECCO 2015, pp. 1479–1480 (2015). https://doi.org/10.1145/2739482.2764693
26. Baioletti, M., Milani, A., Santucci, V.: A new precedence-based ant colony optimization for permutation problems. In: Shi, Y., et al. (eds.) SEAL 2017. LNCS, vol. 10593, pp. 960–971. Springer, Cham (2017). https://doi.org/10.1007/978-3-319-68759-9_79

An Evolutionary Algorithm
with Practitioner's-Knowledge-Based
Operators for the Inventory
Routing Problem

Piotr Lipinski[1(✉)] and Krzysztof Michalak[2]

[1] Computational Intelligence Research Group, Institute of Computer Science,
University of Wroclaw, Wroclaw, Poland
lipinski@cs.uni.wroc.pl
[2] Department of Information Technologies, Institute of Business Informatics,
Wroclaw University of Economics, Wroclaw, Poland
krzysztof.michalak@ue.wroc.pl

Abstract. This paper concerns the Inventory Routing Problem (IRP) which is an optimization problem addressing the optimization of transportation routes and the inventory levels at the same time. The IRP is notable for its difficulty - even finding feasible initial solutions poses a significant problem.

In this paper an evolutionary algorithm is proposed that uses approaches to solution construction and modification utilized by practitioners in the field. The population for the EA is initialized starting from a base solution which in this paper is generated by a heuristic, but can as well be a solution provided by a domain expert. Subsequently, feasibility-preserving moves are used to generate the initial population. In the paper dedicated recombination and mutation operators are proposed which aim at generating new solutions without loosing feasibility. In order to reduce the search space, solutions in the presented EA are encoded as lists of routes with the quantities to be delivered determined by a supplying policy.

The presented work is a step towards utilizing domain knowledge in evolutionary computation. The EA presented in this paper employs mechanisms of solution initialization capable of generating a set of feasible initial solutions of the IRP in a reasonable time. Presented operators generate new feasible solutions effectively without requiring a repair mechanism.

Keywords: Inventory Routing Problem
Dedicated genetic operators · Knowledge-based optimization
Constrained optimization

1 Introduction

The Inventory Routing Problem (IRP) is an extension of the Vehicle Routing Problem (VRP) in which routing optimization is performed jointly with

© Springer International Publishing AG, part of Springer Nature 2018
A. Liefooghe and M. López-Ibáñez (Eds.): EvoCOP 2018, LNCS 10782, pp. 146–157, 2018.
https://doi.org/10.1007/978-3-319-77449-7_10

inventory management optimization [1–3]. A solution of the IRP is a schedule for a planning horizon of T days for a distribution of a single product provided by a single supplier to a number of retailers. The supplier produces a given quantity of the product each day and the retailers sell varying quantities of this product. Both at the supplier and the retailers a limited storage to the product units is available at a cost per unit per day varying from location to location.

The route-optimization part of the IRP is essentially a Vehicle Routing Problem (VRP) [4], because an optimal set of routes has to be found for a fleet of vehicles delivering goods or services to various locations. In line with the VRP each vehicle has a limited capacity and must supply goods to a number of locations satisfying demands of the retailers. The costs of travelling between locations are typically provided in the form of a cost matrix and the total transportation cost is calculated based on all the routes covered by the vehicles. The objective in the VRP is to optimize the transportation cost for the entire fleet [5] and, similarly, in the IRP the transportation cost is calculated for all the vehicles together. Contrary to the VRP the daily demands of the retailers are not fixed and the optimization algorithm has to decide what number of units to deliver each day to satisfy the minimum required inventory level of all the retailers not exceeding the available storage space limits. Also, the cost function represents jointly the costs of storage and transportation so different solutions may trade off one at the expense of the other.

The VRP, as a generalization of the Travelling Salesman Problem (TSP), is an NP-hard problem, and, naturally, so is the IRP. One of the difficulties that distinguish the IRP and the VRP from the TSP is the presence of multiple constraints. In the context of metaheuristic methods this fact necessitates using techniques that ensure feasibility of solutions, such as repair procedures or feasibility-preserving operators.

In the literature various extensions of the regular IRP are defined and studied which arise in real-life applications. For example, in paper [6] the Inventory-Routing Problem with Transshipment (IRPT) was introduced. In this variant of the problem it is allowed to move goods from one retailer to another - a possibility which is useful in the case of numerous sales points of the same retailer. In real life the demands cannot be predicted exactly introducing non-determinism to the problem. Paper [7] introduced the Stochastic Inventory Routing Problem (SIRP) in the context of designing a logistics system for collecting infectious medical wastes. In a location IRP not only are the routes and deliveries optimized, but also the locations of warehouses [8].

Approaches used for solving the IRP include formulating the IRP as an integer programming problem and solving it using methods such as the branch-and-cut algorithm [9]. In papers which tackle real-life applications it is common to use heuristic approaches which use practitioners' experience to construct acceptable solutions. Some well-known heuristics include starting with a solution that serves the retailers with small inventories (who, because of a small storage space available and large sells have to be served every day) using separate vehicles. Such a solution is subsequently modified by adding the remaining retailers, swapping retailers between vehicles, decreasing/increasing the amount of product to deliver, etc. Other heuristics construct solutions by determining, for each retailer, the latest day when it must be supplied to avoid shortage of the product in its inventory and then progressively move the deliveries to earlier dates in order not to overload the vehicles. Paper [10] presents a comparison of a number of heuristics for the IRP, focusing especially on replenishment strategies, but also studying the IRP in the context of an integrated Production Inventory Distribution Routing Problem (PIDRP). Metaheuristic approaches to the IRP include evolutionary algorithms [8], hybrid methods, for example combining simulated annealing and direct search [11] and tabu search [12]. For stochastic optimization problems a common approach is to use simheuristics - hybrid algorithms combining simulation and heuristics [13]. This approach for the stochastic version of the IRP was used in [14].

2   Problem Definition

In this paper, we consider the IRP concerning delivering a single product from a supplier facility S to a given number n of retailer facilities $R_1, R_2, \ldots, R_n$ by a fleet of v vehicles of a fixed capacity C. The supplier S produces p_0 items of the product each day. Each retailer R_i, for $i = 1, 2, \ldots, n$, sells p_i items of the product each day. The supplier has a local inventory, where the product may be stored, with an initial level of $l_0^{(init)}$ items at the date $t = 0$ and with lower and upper limits for the inventory level equal to $l_0^{(min)}$ and $l_0^{(max)}$, respectively. Storing the product in the supplier inventory is charged with an inventory cost c_0 per item per day. Similarly, each retailer R_i, for $i = 1, 2, \ldots, n$, has a local inventory, where the product may be stored, with an initial level of $l_i^{(init)}$ items at the date $t = 0$ and with lower and upper limits for the inventory level equal to $l_i^{(min)}$ and $l_i^{(max)}$, respectively. Storing the product in the retailer inventory is charged with an inventory cost c_i per item per day. The IRP aims at determining the plan of supplying the retailers minimizing the total cost, i.e. for a given planning horizon T, for each date $t = 1, 2, \ldots, T$, the retailers to supply at the date t must be chosen, an amount of the product to deliver to each of these retailers must be determined, and the route of each supplying vehicle must be defined. Formally, the solution is a pair $(\mathbf{R}, \mathbf{Q})$, where $\mathbf{R} = (\mathbf{r}_1, \mathbf{r}_2, \ldots, \mathbf{r}_T)$ is a list of routes in the successive dates $t = 1, 2, \ldots, T$ (each route is a permutation of a certain subset of retailers), and $\mathbf{Q} \in \mathbb{R}^{n \times T}$ is a matrix of column vectors $\mathbf{q}_1, \mathbf{q}_2, \ldots, \mathbf{q}_T$ defining the quantities to deliver to each retailer in the successive dates $t = 1, 2, \ldots, T$ (if a retailer is not included in the route at the date t, the corresponding quantity encoded in the vector $\mathbf{r}_t$ equals 0). The cost of the solution is the sum of the inventory costs and the transportation costs, i.e.

$$\text{cost(solution)} = \sum_{t=1}^{T+1} (l_0^t \cdot c_0 + \sum_{i=1}^{n} l_i^t \cdot c_i) + \sum_{t=1}^{T} \text{transportation-cost}_t, \quad (1)$$

where l_0^t denotes the inventory level of the supplier S at the date t, l_i^t denotes the inventory level of the retailer R_i at the date t, and transportation-cost$_t$ denotes the transportation costs for the supplying vehicle at the date t. The transportation cost is determined by the route of the vehicles and a given distance matrix defining the transportation costs between each two facilities.

An example of an IRP instance, with $n = 10$ retailers, the planning horizon $T = 3$, and a fleet of one vehicle, as well as the optimal solution, is presented in Table 1 and Fig. 1. Table 1 contains the details on lower and upper limits for the inventory level, the inventory costs, the amount of the daily production at the supplier facility, the amount of the daily consumption at the retailers facilities, and the level of inventories at the successive dates of the planning horizon for the optimal solution. Figure 1 (a) presents the location of the facilities. Figure 1 (b)–(d) presents the routes for the successive dates of the planning horizon. The inventory cost for the successive dates is 76.4, 76.47, 76.52, and 75.98. The transportation cost for successive dates is 531, 1237, 94. Therefore, the total solution cost is 2167.37.

Table 1. Illustration of the definition of the IRP - levels of inventories

	S	R_1	R_2	R_3	R_4	R_5	R_6	R_7	R_8	R_9	R_{10}
Min inv. level	0	0	0	0	0	0	0	0	0	0	0
Max inv. level	-	174	28	258	150	126	138	237	129	154	189
Inv. cost	0.03	0.02	0.03	0.03	0.02	0.02	0.03	0.04	0.04	0.02	0.04
Production	635	-	-	-	-	-	-	-	-	-	-
Consumption	-	87	14	86	75	42	69	79	43	77	63
Inv. at $t = 0$	1583	87	14	172	75	84	69	158	86	77	126
Inv. at $t = 1$	2003	0	0	86	75	42	0	79	43	77	126
Inv. at $t = 2$	1721	87	14	172	0	84	69	158	86	77	63
Inv. at $t = 3$	2206	0	0	86	75	42	0	79	43	0	0

3 Evolutionary Approach

In this paper, we propose an evolutionary approach to solving the IRP based on an evolutionary algorithm with dedicated operators based on the knowledge of practitioners in the field. As even simple instances of the IRP are difficult to solve with regular heuristic search methods without additional knowledge on supplying policies, routing strategies, etc. (in many cases, even generating feasible solutions for the initial population is a challenge), the proposed approach uses some popular practitioner techniques to generate feasible solutions first (for the initial population) and to transform solutions without breaking their feasibility (in the mutation operators).

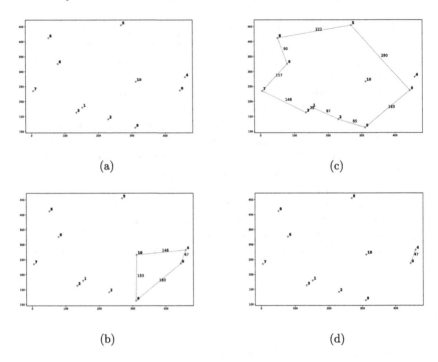

Fig. 1. Illustration of the definition of the IRP - routes in the optimal solution

Algorithm 1 presents the framework of the Evolutionary Algorithm with Practitioner's Knowledge Operators for Inventory Routing Problem (EA-PKO-IRP). It generates an initial population P_1 and evolves it during τ iterations. In each iteration, the current population P_t is evaluated, the offspring population P'_t is created, and the next population P_{t+1} is selected from the union of P_t and P'_t.

Algorithm 1. EA-PKO-IRP

$P_1 = $ Initial-Population(N)
for $t = 1 \rightarrow \tau$ do
    Evaluate(P_t)
    $P'_t = \emptyset$
    for $k = 1 \rightarrow M$ do
        Parent-Solutions = Parent-Selection(P_t)
        Offpring-Solution = Recombination(Parent-Solutions)
        Offpring-Solution = Date-Changing-Mutation(Offpring-Solution)
        Offpring-Solution = Order-Changing-Mutation(Offpring-Solution)
        $P'_t = P'_t \cup \{$Offpring-Solution$\}$
    end for
    $P_{t+1} = $ Replacement($P_t \cup P'_t$)
end for

3.1 Search Space and Solution Encoding

As described in Sect. 2, a solution to the IRP is a pair $(\mathbf{R}, \mathbf{Q})$ consisting of a list of routes $\mathbf{R} = (\mathbf{r}_1, \mathbf{r}_2, \ldots, \mathbf{r}_T)$ for each date and the quantities $\mathbf{Q} = [\mathbf{q}_1, \mathbf{q}_2, \ldots, \mathbf{q}_T]$ to deliver to each retailer in each date of the planning horizon. In this paper, as in many practical approaches, the quantities to deliver are determined by a general supplying policy and are not defined individually, therefore the candidate solution in the evolutionary algorithm is the list of routes $\mathbf{R}$ only. The quantities $\mathbf{Q}$ are defined by a supplying policy, *the up-to-level supplying policy*, that assumes that each retailer is always supplied up to the upper level of its inventory (or not supplied at all, if it is not included in the route of any vehicle for the considered date). Certainly, the supplying policy may limit the IRP problem, but it is frequently used in solving the IRP and usually succeeds in providing efficient solutions.

3.2 Initial Population

The initial population is defined on the basis of a base solution. The base solution is constructed according to a strategy commonly used in practice that tries to supply each retailer at the latest date before the shortage of its inventory. The initial population consists of mutated copies of the base solution.

The base solution is constructed in the following manner: For each date $t = 1, 2, \ldots, T$, a set $\mathcal{R}_t$ of retailers that must be supplied at the date t to avoid the shortage of its inventory at the next date $t+1$ is determined. The quantities to deliver are determined according to *the up-to-level supplying policy*, i.e. the retailer is always supplied up to the upper level of its inventory. The routes of the vehicles are determined in a greedy manner: Each retailer R from the set $\mathcal{R}_t$ is considered in turn (in a random order). For each vehicle $j = 1, 2, \ldots, v$, an attempt to add the retailer R to the route of the vehicle j is made, if the total capacity of the vehicle does not exceed the maximum capacity. The retailer R is added between the supplier node and the first node on the route and the transportation cost is evaluated. Then, the retailer R is shifted between the first and the second node on the original route and the transportation cost is evaluated, etc. Finally, the retailer R is assigned to the vehicle and to the position on the route of the vehicle that has the minimal transportation cost. It may happen that there are no vehicles to consider, because all the vehicles are overloaded. Then, the strategy tries to shift the retailer to an earlier date and find, in a similar manner, a route to add it.

3.3 Recombination Operator

The recombination operator takes T parent solutions $\mathbf{R}^{(1)}, \mathbf{R}^{(2)}, \ldots, \mathbf{R}^{(T)}$, where T is the planning horizon, and produces one offspring solution $\tilde{\mathbf{R}}$ in such a way that

$$\tilde{\mathbf{r}}_i = \mathbf{r}_i^{(\pi_i)}, \qquad \text{for } i = 1, 2, \ldots, T, \tag{2}$$

where $\pi = (\pi_1, \pi_2, \ldots, \pi_T)$ is a random permutation of the indices $1, 2, \ldots, T$ of the parent solutions $\mathbf{R}^{(1)}, \mathbf{R}^{(2)}, \ldots, \mathbf{R}^{(T)}$. If such an offspring solution is not feasible, the procedure is repeated anew (with a different permutation π), up to κ_R times (κ_R is a constant parameter of the algorithm), otherwise the offspring solution is a copy of a parent solution randomly chosen with the same probability of being chosen for each parent equal to $1/T$.

It is worth noticing that at the beginning of the algorithm, when the candidate solutions in the population are usually very different, many produced offspring solutions are infeasible, because the parts of different parent solutions are usually contradictory and cannot be combined into a feasible solution. However, when the population becomes more homogeneous, the parts of parent solutions are usually similar, so many produced offspring solutions are feasible.

3.4   Date-Changing Mutation (DM)

The mutation operators takes one solution $\mathbf{R}$ and modifies it in the following manner: First, a date t is randomly chosen with the uniform distribution over the dates $2, 3, \ldots, T$. Next, a retailer R is randomly chosen from the retailers assigned to service at the date t, i.e. from the route $\mathbf{r}_t$. The retailer R is removed from the route $\mathbf{r}_t$ and all the routes for all the further dates. Next, a date t' is randomly chosen with the uniform distribution over the dates $1, 2, \ldots, t-1$. The retailer R is assigned to service at the date t' and added to the route $\mathbf{r}_{t'}$ in a greedy manner, as in creating the base solution described in Sect. 3.2. Similarly to creating the base solution, the further latest dates when the retailer R must be supplied to avoid the shortage of its inventory are determined, and the retailer R is added to the proper routes. If such a modified solution is not feasible, the procedure is repeated anew, up to κ_M times (κ_M is a constant parameter of the algorithm), otherwise the original solution remains unchanged.

It is worth noticing that the mutation operator changes the schedule concerning the only one selected retailer and always leaves the other retailers untouched. In addition, the mutation operator does not change the schedule before the selected date t' and leaves the beginning of the schedule unchanged.

3.5   Order-Changing Mutation (OM)

The Order-Changing Mutation (OM) operator takes one solution $\mathbf{R}$ and aims at optimizing the routes without changing the assignment of the retailers to the routes. It analyzes each route $\mathbf{r}_1, \mathbf{r}_2, \ldots, \mathbf{r}_T$ and tries to change the order of the retailers on the route. For short routes of no more than ρ retailers (ρ is a constant parameter of the algorithm), each permutation of the retailers is evaluated. For longer routes, $\rho!$ random permutations of the retailers are evaluated. If an evaluated route outperforms the original one, the original route is replaced with the best found alternative.

It is worth noticing that the OM operator does not change the dates of supplying the retailers, so it does not affect the feasibility of the solution.

4   Experiments

The experiments presented in this paper were performed using benchmark IRP instances, published in [9], with the planning horizon T of 3 days, with 5, 10, 15 or 20 retailers, with the inventory costs between 0.01 and 0.05, and with different locations of the facilities and various inventory, production and consumption levels. All the benchmark instances concern a fleet of one vehicle. Table 2 presents the list of the benchmark IRP instances used in the experiments.

Table 2. List of benchmark IRP instances used in the experiments

$n = 5$	5 instances with the planning horizon $T = 3$ and the inventory costs between 0.01 and 0.05
$n = 10$	5 instances with the planning horizon $T = 3$ and the inventory costs between 0.01 and 0.05
$n = 15$	5 instances with the planning horizon $T = 3$ and the inventory costs between 0.01 and 0.05
$n = 20$	5 instances with the planning horizon $T = 3$ and the inventory costs between 0.01 and 0.05

For each benchmark instance, the proposed EA-PKO-IRP algorithm was run 10 times in order to reduce the influence of the randomness of the algorithm on the presented results. A number of different parameter settings were investigated, taking into account the efficiency of the algorithm as well as the computing time on a few selected problem instances, and the optimal parameter setting was used in all the experiments. Table 3 presents the parameter settings of the EA-PKO-IRP algorithm.

Table 3. Parameter settings of the EA-PKO-IRP algorithm

Description	Symbol	Value
Population size	N	500
Number of offspring solutions	M	2000
Number of parents for each offspring	k	T
Number of iterations	τ	100
Replacement parameter	κ_R	10
DM mutation parameter	κ_M	5
OM mutation parameter	ρ	6

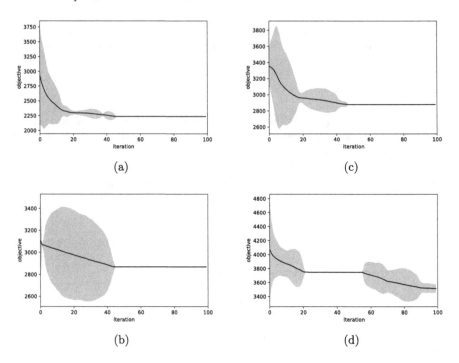

Fig. 2. Evolution of the values of the objective function in the successive iterations of the evolutionary algorithm for 4 selected cases (the gray area shows the standard deviation of the objective function value in the population)

Figure 2 presents the evolution of the values of the objective function in the successive iterations of the evolutionary algorithm for 4 selected cases (the gray area shows the standard deviation of the objective function value in the population). Figure 2 (a) presents the typical behavior – the diversity of the population is large at the beginning and narrows down in successive iterations. Figures 2 (b), (c), (d) present some interesting behaviors: (b) – the diversity of the population is increasing at the beginning to explore the search space (perhaps, the initial population was not diversified enough); (c) – a similar effect occurs after about 20 iterations (perhaps, after converging to a local minimum); (d) – after converging to a local minimum after about 20 iterations, the algorithm is trying to find a better solution, but recombinations probably lead to infeasible solutions, and mutations cannot improve the local minimum.

Table 4 presents the results of the proposed EA-PKO-IRP algorithm on 20 benchmark IRP instances. The first column contains the name of the benchmark, published in [9]. The second column recalls the exact optimum, published in [9].

Table 4. Results of EA-PKO-IRP on 20 benchmark IRP instances. f_{opt} denotes the optimal value obtained using exact methods presented in [9].

Benchmark	Optimum (f_{opt})	Best of 10 runs (f_b)	Mean of 10 runs (f_m)	$f_b - f_{opt}$	$f_m - f_{opt}$
abs1n5	1281.68000	1281.68000	1281.68000	0.00000	0.00000
abs2n5	1176.63000	1176.63000	1176.63000	0.00000	0.00000
abs3n5	2020.65000	2020.65000	2020.65000	0.00000	0.00000
abs4n5	1449.43000	1449.43000	1449.43000	0.00000	0.00000
abs5n5	1165.40000	1165.40000	1165.40000	0.00000	0.00000
abs1n10	2167.36999	2167.37000	2167.37000	0.00000	0.00000
abs2n10	2510.12988	2510.13000	2510.13000	0.00010	0.00010
abs3n10	2099.67993	2099.68000	2099.68000	0.00010	0.00010
abs4n10	2188.00999	2188.01000	2190.51000	0.00000	2.50000
abs5n10	2178.15000	2178.15000	2178.15000	0.00000	0.00000
abs1n15	2236.52999	2236.53000	2236.53000	0.00000	0.00000
abs2n15	2506.20996	2506.21000	2506.21000	0.00000	0.00000
abs3n15	2841.05999	2841.06000	2854.26000	0.00000	13.20000
abs4n15	2430.06999	2430.07000	2439.44400	0.00000	9.37400
abs5n15	2453.49999	2453.50000	2464.03900	0.00000	10.53900
abs1n20	2793.28999	2879.56000	2879.56000	86.27000	86.27000
abs2n20	2799.89999	2867.89000	2877.10000	67.99000	77.20000
abs3n20	3101.59999	3950.80000	3950.80000	849.20000	849.20000
abs4n20	3239.30999	3322.65000	3492.97300	83.34000	253.66300
abs5n20	3330.98999	3396.98000	3452.91900	65.99000	121.92900

The next two columns present the best and the mean result of the 10 runs of the proposed EA-PKO-IRP algorithm. The last two columns present the difference between the results and the exact optimum.

Table 5 presents the results of the proposed EA-PKO-IRP algorithm on 20 benchmark IRP instances after additional optimization of routes. The additional optimization of routes is a post-processing technique applied to the best candidate solution found by the algorithm after the evolution has terminated. In this step the order-changing mutation operator (OM) is applied to the best solution found by the evolutionary algorithm.

Table 5. Results of EA-PKO-IRP on 20 benchmark IRP instances after additional optimization of routes. f_{opt} denotes the optimal value obtained using exact methods presented in [9].

Benchmark	Optimum (f_{opt})	Best of 10 runs (f_b)	Mean of 10 runs (f_m)	$f_b - f_{opt}$	$f_m - f_{opt}$
abs1n5	1281.68000	1281.68000	1281.68000	0.00000	0.00000
abs2n5	1176.63000	1176.63000	1176.63000	0.00000	0.00000
abs3n5	2020.65000	2020.65000	2020.65000	0.00000	0.00000
abs4n5	1449.43000	1449.43000	1449.43000	0.00000	0.00000
abs5n5	1165.40000	1165.40000	1165.40000	0.00000	0.00000
abs1n10	2167.36999	2167.37000	2167.37000	0.00000	0.00000
abs2n10	2510.12988	2510.13000	2510.13000	0.00010	0.00010
abs3n10	2099.67993	2099.68000	2099.68000	0.00010	0.00010
abs4n10	2188.00999	2188.01000	2188.01000	0.00000	0.00000
abs5n10	2178.15000	2178.15000	2178.15000	0.00000	0.00000
abs1n15	2236.52999	2236.53000	2236.53000	0.00000	0.00000
abs2n15	2506.20996	2506.21000	2506.21000	0.00000	0.00000
abs3n15	2841.05999	2841.06000	2854.26000	0.00000	13.20000
abs4n15	2430.06999	2430.07000	2439.44400	0.00000	9.37400
abs5n15	2453.49999	2453.50000	2464.03900	0.00000	10.53900
abs1n20	2793.28999	2879.56000	2879.56000	86.27000	86.27000
abs2n20	2799.89999	2867.89000	2877.10000	67.99000	77.20000
abs3n20	3101.59999	3905.80000	3905.80000	804.20000	804.20000
abs4n20	3239.30999	3322.65000	3488.87300	83.34000	249.56300
abs5n20	3330.98999	3396.98000	3449.11900	65.99000	118.12900

5   Conclusions

In this paper an evolutionary algorithm for the Inventory Routing Problem (IRP) is presented. Because of the intricacy of the IRP the proposed method employs numerous mechanisms which, by utilizing the experience of practitioners in the field, reduce the complexity of the task faced by the evolutionary optimizer. The population initialization procedure starts with the procedure commonly used for generating feasible solutions of the IRP. Subsequently, mutation operators are used to obtain a diversified initial population fast. The operators proposed in the paper retain the feasibility of solutions, thereby obviating the need for a repair procedure. The work presented in this paper is a step towards utilizing domain knowledge and good practices in evolutionary computation. Further work may concern generalizing the presented approach to a wider range of optimization problems.

Acknowledgment. This work was supported by the Polish National Science Centre (NCN) under grant no. 2015/19/D/HS4/02574. Calculations have been carried out using resources provided by Wroclaw Centre for Networking and Supercomputing (http://wcss.pl), grant no. 405.

References

1. Aghezzaf, E.H., Raa, B., Van Landeghem, H.: Modeling inventory routing problems in supply chains of high consumption products. Eur. J. Oper. Res. **169**(3), 1048–1063 (2006)
2. Archetti, C., Bianchessi, N., Irnich, S., Speranza, M.G.: Formulations for an inventory routing problem. Int. Trans. Oper. Res. **21**(3), 353–374 (2014)
3. Bertazzi, L., Speranza, M.G.: Inventory routing problems: an introduction. EURO J. Transp. Logist. **1**(4), 307–326 (2012)
4. Dantzig, G.B., Ramser, J.H.: The truck dispatching problem. Manag. Sci. **6**(1), 80–91 (1959)
5. Laporte, G.: Fifty years of vehicle routing. Transp. Sci. **43**(4), 408–416 (2009)
6. Coelho, L.C., Cordeau, J.F., Laporte, G.: The inventory-routing problem with transshipment. Comput. Oper. Res. **39**(11), 2537–2548 (2012)
7. Nolz, P.C., Absi, N., Feillet, D.: A stochastic inventory routing problem for infectious medical waste collection. Networks **63**(1), 82–95 (2014)
8. Hiassat, A., Diabat, A., Rahwan, I.: A genetic algorithm approach for location-inventory-routing problem with perishable products. J. Manuf. Syst. **42**, 93–103 (2017)
9. Archetti, C., Bertazzi, L., Laporte, G., Speranza, M.G.: A branch-and-cut algorithm for a vendor-managed inventory-routing problem. Transp. Sci. **41**(3), 382–391 (2007)
10. Bard, J.F., Nananukul, N.: Heuristics for a multiperiod inventory routing problem with production decisions. Comput. Ind. Eng. **57**(3), 713–723 (2009)
11. Diabat, A., Dehghani, E., Jabbarzadeh, A.: Incorporating location and inventory decisions into a supply chain design problem with uncertain demands and lead times. J. Manuf. Syst. **43**, 139–149 (2017)
12. Liu, S.C., Chen, J.R.: A heuristic method for the inventory routing and pricing problem in a supply chain. Expert Syst. Appl. **38**(3), 1447–1456 (2011)
13. Juan, A.A., et al.: A review of simheuristics: extending metaheuristics to deal with stochastic combinatorial optimization problems. Oper. Res. Perspect. **2**, 62–72 (2015)
14. Juan, A.A., Grasman, S.E., Caceres-Cruz, J., Bekta, T.: A simheuristic algorithm for the single-period stochastic inventory-routing problem with stock-outs. Simul. Model. Pract. Theor. **46**, 40–52 (2014)

A Multistart Alternating Tabu Search for Commercial Districting

Alex Gliesch[1(✉)], Marcus Ritt[1], and Mayron C. O. Moreira[2]

[1] Federal University of Rio Grande do Sul, Porto Alegre, Brazil
{alex.gliesch,marcus.ritt}@inf.ufrgs.br
[2] Federal University of Lavras, Lavras, Brazil
mayron.moreira@dcc.ufla.br

Abstract. In this paper we address a class of commercial districting problems that arises in the context of the distribution of goods. The problem aims at partitioning an area of distribution, which is modeled as an embedded planar graph, into connected components, called districts. Districts are required to be mutually balanced with respect to node attributes, such as number of customers, expected demand, and service cost, and as geometrically-compact as possible, by minimizing their Euclidean diameters. To solve this problem, we propose a multistart algorithm that repeatedly constructs solutions greedily and improves them by two alternating tabu searches, one aiming at achieving feasibility through balancing and the other at maximizing district compactness. Computational experiments confirm the effectiveness of the different components of our method and show that it significantly outperforms the current state of the art, improving known upper bounds in almost all instances.

Keywords: Districting · Territory design · Tabu search
Heuristic algorithm · Compactness

1 Introduction

The goal of districting problems is to group basic geographic units into clusters, called districts, which satisfy given constraints. Typically, units represent geographic entities such as city blocks, and have some attributes, such as population or expected travel cost. An underlying planar graph establishes adjacencies between these units. Common requirements include that districts be equally balanced with respect to attribute values, be geometrically compact by having a more or less regular convex shape, and be connected with respect to unit adjacencies.

In recent years, districting problems have been studied in a wide range of applications, such as the design of electoral districts [1–4], sales territories [5–7], police districts [8], health care districts [9,10], or agrarian land parcels [11]. [12] provides an extensive overview on applications and solution techniques to many districting problems.

© Springer International Publishing AG, part of Springer Nature 2018
A. Liefooghe and M. López-Ibáñez (Eds.): EvoCOP 2018, LNCS 10782, pp. 158–173, 2018.
https://doi.org/10.1007/978-3-319-77449-7_11

In this paper, we study districting applied to a commercial territory design problem that originates from a real-world context of a bottled beverage distribution company. In this problem, each geographic unit represents a city block and has three attributes, or "activities": the number of customers, the product demand, and the workload. The goal is to divide the city blocks into a fixed number of contiguous, compact districts that are balanced with respect to all three activities.

The problem's domain was first introduced by [5]. They present a core model that optimizes compactness through an objective function based on the dispersion measure of the well-known p-centers problem, while treating contiguity and balance of activities as constraints. They further show that the problem is NP-hard and propose heuristic solutions by a reactive GRASP strategy.

Since then, several models and solution approaches to this problem have been studied. [13] focus on solving it optimally through mixed-integer programming. They also propose a variant model whose objective function is based on the p-median problem, as opposed to a p-center approach (i.e. minimizing maximum and not average distance to the centers). [14] study a multi-objective variant that optimizes both compactness (through a p-median approach) and balance. More recently, [7] proposed a GRASP with a path-relinking strategy to solve a novel variant that needs no district centers for compactness, as the p-center and p-median approaches do, but uses the more flexible concept of polygon diameters.

In this paper, we address the more recent model by [7]. Section 2 presents the problem formally and provides the necessary definitions. In Sect. 3 we propose a new method for solving this problem, using a multistart tabu search, which alternates between improving balance constraints and maximizing compactness. In Sect. 4, we report on the results of computational experiments which evaluate the different components of the proposed method, and compare the quality of the solutions to approaches from the literature. We conclude in Sect. 5.

2   Definitions

We are given an undirected connected planar graph $G = (V, E)$ of $n = |V|$ nodes and $m = |E|$ edges, and a number of districts $p \leq n$. Each node $u \in V$ corresponds to a basic geographic unit and is associated with three activity values w_{u1}, w_{u2}, w_{u3} and coordinates $x_u \in \mathbb{R}^2$. In the following, we use the set notation $[n] = \{1, \ldots, n\}$.

A districting plan (or solution) S for G is a partition of a subset of the vertices $S_1 \dot\cup \cdots \dot\cup S_p \subseteq V$, where each S_i represents a district. We say that a solution S is *complete* if $\bigcup_{i \in [p]} S_i = V$. We use the notation $S(u) = i$ to mean that node u is *assigned to* district i, i.e., $u \in S_i$. If S is incomplete, then some $S(u)$ will be undefined, in which case we say that u is *unassigned*.

We define the *boundary* of a district S_i as $\partial S_i = \{v \mid \{v, u\} \in E, u \in S_i, v \notin S_i\}$, and its free boundary as $\partial_f S_i = \{v \in \partial S_i \mid v \text{ is unassigned}\}$. We say that districts i and j are *neighbors* if $S_i \cap \partial S_j \neq \emptyset$. A *move* is an operation $u \to i$ that shifts the assignment of node u to district i. We use the notation $S[\nu]$ to refer to S after performing move ν.

Let $w_a(S_i) = \sum_{u \in S_i} w_{ua}$ be the total value of district i with respect to one of the three activities $a \in [3]$, and μ_a be the mean value of w_{ua} over all $u \in V$. We define $b_a(S_i) = |(w_a(S_i) - \mu_a)/\mu_a|$ as the absolute relative deviation of the total value of activity a from the mean activity a among all districts. We say that a district S_i is *balanced* if $b_a(S_i) \leq \tau_a$ for $a \in [3]$, where $\tau_a \geq 0$ is a tolerance parameter. A solution is balanced if all its districts are balanced. The *imbalance* of a district S_i is defined as the excess of the activities over the tolerances $B(S_i) = \sum_{a \in [3]} \max\{0, b_a(S_i) - \tau_a\}$, and the imbalance of a solution S is the total excess of all districts $B(S) = \sum_{i \in [p]} B(S_i)$. A solution S is balanced iff $B(S) = 0$.

Finally, we define the *diameter* of a district S_i as $D(S_i) = \max_{u,v \in S_i} d_{uv}$, where $d_{uv} = |x_u - x_v|_2$ is the Euclidean distance between nodes u and v, and the diameter of a solution S as the maximum diameter $D(S) = \max_{i \in [p]} D(S_i)$. Since the diameter is used here as a measure for compactness, where a small diameter corresponds to a high compactness, we will use both terms interchangeably.

A districting plan S is considered feasible if (1) it is complete, (2) it is balanced, and (3) the subgraph induced by each S_i in G is connected. Our goal is to find a feasible districting plan that minimizes $D(S)$.

3   Proposed Methods

The overall structure of our algorithm is similar to a greedy randomized adaptive search procedure (GRASP, [15]): we repeatedly construct a solution using a randomized greedy algorithm and improve it by a heuristic based on local search, returning the best solution found over all repetitions. Our method differs from traditional GRASP in two ways. First, during construction we do not select candidates randomly from a restricted candidate list, as is standard, but select the best overall candidate in two stages by different criteria. Second, instead of improving solutions by local search, we perform two tabu searches alternatedly, one focusing on making solutions feasible by reducing imbalance and the second one focusing on optimizing compactness.

Using different searches for each objective has been applied sucessfully to districting before [6,16]. The main idea is to focus the effort on optimizing one objective by searching a custom neighborhood that only considers moves with respect to that objective, while (to some extent) ignoring others. A common alternative would be to perform a single search using a composite fitness function that assigns weights to each objective, as is done by [3] and [7]. The main disadvantage of this approach is that it is usually difficult to find weights that work well for all instances.

Algorithm 1 outlines our method. Each iteration starts by computing an initial solution through a greedy constructive algorithm (line 3). Next, the local improvement phase (lines 5–10) optimizes first compactness and then balance, repeating this for a fixed number of iterations $A_{\max}$. After the first iteration,

Algorithm 1. Multistart alternating tabu search.
```
 1: R ← ∅
 2: repeat
 3:     S ← greedyRandomizedConstruction()
 4:     X = {S}
 5:     for i ∈ [A_max] do
 6:         S ← optimizeCompactness(S)
 7:         S ← optimizeBalance(S, α)
 8:         if S ∈ X then
 9:             break
10:         X ← X ∪ {S}
11:     S ← arg min{(B(S′), D(S′)) | S′ ∈ X}
12:     S ← optimizeBalance(S, ∞)
13:     R ← arg min{B(R), B(S)}
14: until stopping criterion satisfied
15: return R
```

subsequent searches are typically much faster, since solutions are already close to a local minimum in both criteria. When optimizing for balance we do not allow the diameter of the solution found in the preceding optimization of compactness to increase by more than a factor α, where $\alpha \geq 0$ is a slack parameter. This maintains a certain progress in the reduction of the diameter. We store intermediate solutions in a lookup table after each iteration and stop if cycling is detected (lines 8–9). This is done with a hash table; since A_{max} is expected to be small (less than 1000), this has is no effect on performance. In the end, we select the intermediate solution with minimum D, or the one with minimum B if no balanced solutions were found (line 11). In the latter case, we make a final attempt at finding a feasible solution by optimizing for balance with $\alpha = \infty$ (line 12). We iterate until a stopping criterion (either a maximum number of iterations or a time limit) is met, and report the best solution found.

A tabu search is used to optimize compactness and balance (procedures optimizeCompactness and optimizeBalance). Tabu search is a non-monotone local search proposed by [17]. Starting from an initial solution, it repeatedly moves it to the best neighbor. To avoid cycling, some neighbors are declared tabu, i.e. they cannot be selected. The two tabu searches in our algorithm explore different neighborhoods, iteratively performing local movements for at most I_{max} consecutive non-improving iterations, and returning the best intermediate solution. The tabu list contains nodes whose assignment was recently shifted and prohibits to shift them again for t iterations, where the tabu tenure t is a parameter. The tabu list is emptied between consecutive searches.

In the following subsections, we describe each method (greedy construction and tabu searches) in more detail.

3.1   Solution Construction

Initial solutions are constructed in two steps: first, we select p initial nodes which will serve as seeds for each district, and then we grow districts by iteratively assigning them a greedily-selected node on their free boundary, until the solution is complete.

In the first step, we compute seed nodes s_i for each district $i \in [p]$. Our method is based on the greedy constructive heuristic proposed by [18] to solve the p-dispersion problem. The p-dispersion problem aims at finding p elements from a set of points on the plane such that the minimum distance between any two elements is maximized. First, we select s_1 randomly from V. Then, for each $i \in [2, p]$ in sequence, we pick a subset $V' \subseteq V \setminus \{s_1, \ldots, s_{i-1}\}$ of size $\sqrt{n}$ uniformly at random, and define the ith center $s_i = \arg \max_{u \in V'} \min_{j \in [i-1]} d_{us_j}$ to be the node in V' that maximizes the minimum distance to the previously chosen centers. [7] use a similar strategy, with the difference that each s_i is chosen from $V' = V \setminus \{s_1, \ldots, s_{i-1}\}$. The main reason we chose to reduce V' to a smaller, random subset is that, with a time complexity of $\Theta(p^2|V'|)$, the former method happens to be too slow for larger instances. Our approach also has the advantage of increasing variability. In practice, we have found the two strategies to have very little difference in average solution quality.

The second step starts from districts $S_i = \{s_i\}$. We maintain for each district i the candidate node $c_i \in \partial_f S_i$ that minimizes $D(S_i \cup \{c_i\})$. If multiple such nodes exist, we select one randomly. At first, we compute c_i for every district by iterating over each $\partial_f S_i$. Then, we repeatedly select the district j with smallest total normalized activity $\sum_{a \in [3]} w_a(S_j)/\mu_a$ and assign $S := S[c_j \rightarrow j]$, until S is complete. We recompute c_i for all districts S_i with $c_i = c_j$ after every assignment, because the free boundary $\partial_f S_i$ of these districts changes. There are, in average, only $O(1)$ such districts, since G is planar, and thus has an average degree less than 6. Because only boundary nodes are considered, connectivity is preserved.

This strategy is greedy in the way that it selects a candidate with smallest increase of D for each district. This steers the construction towards initial solutions with low diameters, as opposed to a different strategy that might, for example, try to find solutions that are as balanced as possible by minimizing B. Our rationale is that, in most cases, even highly imbalanced solutions can be made feasible quickly with a local search-based procedure, whereas it is substantially harder to obtain large improvements in compactness. Nonetheless, by selecting the district with minimum activity at each step, we ensure a certain balancing and prevent districts in denser regions from growing too large.

3.2   Optimizing Balance

We optimize for balance by shifting boundary nodes between neighboring districts. More specifically, we consider moves in the neighborhood $N_{bal}(S) = \{u \rightarrow i \mid i \in [p], u \in \partial S_i\}$. As explained above, we avoid moves which lead to a diameter that exceeds the diameter D_0 of the starting solution (obtained after optimizing compactness) by more than a factor α. Thus, the core search algorithm repeatedly selects a move $u \rightarrow i \in N_{bal}$ such that:

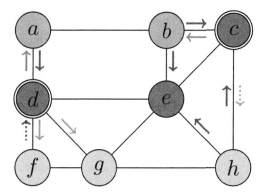

Fig. 1. Neighborhood N_{bal}. Each district is represented by a color. Colored arrows emanating from a node represent a possible shift of the node to the district of that color. We have omitted moves that break connectivity, as is the case of all shifts of e and g. The doubly-circled nodes c and d are the farthest same-district nodes, and make up the solution's diameter. The dotted arrows from $c{\to}\square$ and $f{\to}\blacksquare$ represent moves that would increase the current diameter (the new farthest pair would be (c, f) in both cases): depending on the value of α, they might not be applicable. (Color figure online)

1. u is not tabu
2. $D(S[u{\to}i]) \leq (1 + \alpha)D_0$
3. $S(u)$ remains connected after removing u
4. $B(S[u{\to}i])$ is minimized

If multiple such moves exist, we select one of them randomly. We then assign $S := S[u{\to}i]$, and mark u as tabu. We stop either after I_{max} consecutive steps without improvement, or if there are no more valid moves, or if the solution is balanced ($B(S) = 0$), since there can be no further improvement, and return the intermediate solution with minimum $(B(S), D(S))$, compared lexicographically. Figure 1 illustrates the effective neighborhood seached with a minimalistic example.

Similarly to the constructive algorithm of Sect. 3.1, we can avoid re-evaluating all of N_{bal} at every iteration by caching, for each district i, a candidate $c_i \in \partial S_i$ that satisfies items 1, 2, 3 and 4 above. Each c_i is initially computed by iterating over ∂S_i. At every iteration, we choose the candidate c_i, $i \in [p]$, with lowest expected imbalance and perform $S := S[c_i{\to}i]$. Upon assigning $c_i{\to}i$, we update c only for the districts affected by the move (i.e., i and $S(c_i)$) and their direct neighbors. Again, due to G being planar, there are in average only $O(1)$ such districts (here, we can view districts as nodes and their neighboring relations as edges). Further, at each iteration a node ceases to be tabu, and thus we also update c for neighboring districts of that node.

3.3 Optimizing Compactness

Only a small subset of the neighborhood N_{bal} actually reduces the diameter, and so when $B(S) = 0$ or is at a plateau it would be wasteful to continue to search all of N_{bal}. Let the set $L(S) = \{u \mid \exists v \in S, S(u) = S(v), d_{uv} = D(S)\}$ contain all nodes incident to some maximum diameter of S. When optimizing compactness, we consider only moves in the neighborhood $N_{cmp}(S) = \{u \to i \mid i \in [p], u \in L(S) \cap \partial S_i\}$ which have the potential to improve $D(S)$. In the example of Fig. 1, we have $L = \{c, d\}$ and $N_{cmp} = \{d \to \blacksquare, d \to \square, c \to \blacksquare, c \to \square\}$.

Note that if $|L(S)| > 2$, then even shifting a node in L might not change D, since there still exists another pair of nodes with equal distance (except if a node is incident to multiple diameters). This situation happens frequently when the node locations are regularly distributed, for example, in a grid. It is clear, however, that reducing the cardinality of L is still an improvement. Thus, we rank such moves accordingly by using a modified objective function $D'(S) = D(S) + \epsilon |L(S)|$, where ϵ is some small constant. In our implementation, we use $\epsilon = 5 \times 10^{-7}$.

We then optimize for compactness by repeatedly searching for the move $u \to i \in N_{cmp}$ with minimum $D'(S[u \to i])$, and assigning $S := S[u \to i]$. If multiple such moves exist, we select one at random. As before, we discard moves that break connectivity or are tabu, and stop either after I_{max} steps without improvement or if there are no more valid moves, and return the best intermediate solution with respect to $(D(S), B(S))$ compared lexicographically.

A common situation occurs when no nodes in L are on the boundary of another district, meaning that $N_{cmp}(S) = \emptyset$ and thus no moves can be made. This represents a plateau that is particularly difficult to escape, since it requires at least as many moves as the length of the shortest path from a node in L to the boundary of some district. In practice, it is very unlikely that this specific set of moves will be performed while searching N_{bal}. Therefore, when this kind of plateau is reached during search, we attempt to escape it as follows. We search G for the path π of smallest Euclidean distance starting from any node $u \in L(S)$ and ending at some node $v \in \partial S_{S(u)}$. We then assign all nodes in π, including u but not v, to district $S(v)$, as long as $S(u)$ remains connected (otherwise we stop and give up). Reassigning node u forces a change in L, and consequently in D' (barring the rare case that u remains in L, paired to the same number of nodes and distances as before). We repeat this process at most sp_{max} times, where sp_{max} is a parameter, or until $|N_{cmp}| > 0$. Using a standard shortest paths algorithm that orders active nodes with a priority queue, finding path π takes, on average, $O(n/p \log n/p)$ steps. Figure 2 illustrates the process, which call shortest path escape, or sp-escape, for short.

3.4 Data Structures for Efficient Operations

The proposed algorithm depends on repeated tests of connectivity and queries of the diameter of districts. Implemented naively, these operations can become the bottleneck of the method. In this section we explain the data structures we have used to implement them efficiently.

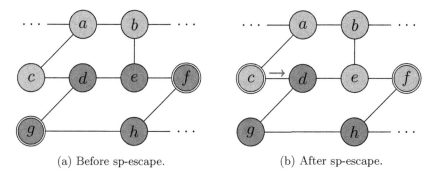

(a) Before sp-escape.   (b) After sp-escape.

Fig. 2. The sp-escape process. We use "$\cdots$" to mean that the solution continues with other districts at that point. In Fig. 2a, the two farthest nodes g and f are not on the boundary of any other district (i.e., $N_{cmp} = \emptyset$). We therefore search for the shortest distance path from either g or f to the boundary of ■, and find feb. We then assign all nodes in feb to ■, obtaining the solution in Fig. 2b. Notice that, after this process, the search is able to continue, since $N_{cmp} = \{c{\rightarrow}■\}$. (Color figure online)

Connectivity queries. Testing for loss of connectivity when considering a move $u{\rightarrow}i$ by performing a graph search on $S(u)$ is very time-consuming, taking $O(n/p)$, on average, and can easily dominate the running time of the algorithm, since a large number of candidates must be evaluated before performing each move.

The size of district boundaries in planar domains in general scales closely to $O(\sqrt{n/p})$. Under this assumption, we can test for loss of connectivity in an amortized time of $O(\sqrt{n/p})$ by maintaining a bit array of size n indicating whether each node's removal will disconnect the district owning it. This array is computed for each district i in time $O(|S_i|)$ using the standard algorithm to find articulation nodes [19]. Before the local improvement phase starts, we compute the articulation nodes of all districts. During search, upon applying a move $u{\rightarrow}i$, we need only to recompute articulation nodes in two districts: $S(u)$ and i. Although this update takes $O(n/p)$ time, on average, when searching N_{bal} this is only done once for every full evaluation of a district boundary, of average size $O(\sqrt{n/p})$.

This could be improved even further by using the algorithm of [20] for decremental connectivity in planar graphs, as it allows both tests and updates in constant time, or the geo-graph model of [21], which provides an efficient method for maintaining connectivity and hole constraints in districting problems.

Maintaining diameters. A brute-force method computes the diameter of a district i in $O(|S_i|^2)$ by checking the distance between all node pairs. Similarly, the expected diameter of a move $u{\rightarrow}i$ can be computed in $O(|S_i|)$ by checking the distance from u to all nodes in S_i.

We can do better by using the fact that the most distant pair of points of a point set must lie on that point set's convex hull [22]. Knowing that the convex hull of a set of n uniformly-distributed points on the plane has expected size $O(\log n)$ [23] and assuming we maintain the convex hull of every district, we can compute the diameter of district i in expected time $O(\log |S_i|)$ by executing the so-called "rotating calipers" algorithm [22] on the convex hull of S_i. (Here, we assume that node coordinates are more or less uniformly-distributed.)

We compute convex hulls with the monotone chain algorithm [24], which runs in $O(n \log n)$ time for any set of n points or in $O(n)$ time if the points are already sorted lexicographically by their coordinates. Thus, if we sort the list of nodes of every district by their coordinates before the local improvement phase and keep them sorted after each move using a simple linear update, we can maintain the convex hulls in time $O(|S_i|)$ per update. This could be further improved using the algorithm of [25], which maintains convex hulls of n points in $O(\log^2 n)$ time per update.

4   Computational Experiments

4.1   Test Instances

[7] report results on two data sets called DS and DT. Both contain 20 randomly generated instances, with each instance having $n = 500$ nodes, $p = 10$ districts and m ranging from 917 to 986 edges. These data sets were originally proposed by [5] and were generated to resemble real-world scenarios. Data set DS selects node activities uniformly from the intervals $[4, 20], [15, 400]$, and $[15, 100]$, and data set DT selects activities from a non-uniform symmetric distribution. We refer to [5] for more details on these data sets.

Because all the instances in these data sets have the same size, for a broader evaluation we have generated an additional data set "DL" with 4 instances of each combination $n \in \{1000, 2500, 5000, 10000\}$ and $p \in \{n/200, n/100, n/62.5\}$, 48 in total. The three levels of p roughly represent difficulties "easy", "medium" and "hard", with respect to achieving feasibility. We have chosen these size ranges because no existing methods were able to consistently find feasible solutions for larger instances, whereas smaller instances tended to be trivial. For DL, we have decided to use the same uniform activity generation used for generating data set DS, as we have found it significantly more challenging than the one used for DT. As in [7], we use balancing tolerances $\tau_1 = \tau_2 = \tau_3 = 0.05$. The coordinates of each node were drawn uniformly at random from $[0, 1000]^2$. The graph topology was obtained by computing a Delaunay triangulation on the node coordinates, which ensures connectivity and planarity.

4.2   Experimental Setup

We have implemented our algorithms in C++ and compiled them with GCC 5.4.0 and maximum optimization. The code, data sets, instance generators and detailed results of the experiments are available to the community upon request. All experiments were performed on a PC with an 8-core AMD FX-8150 processor

Table 1. Parameter calibration: initial ranges and best setting found by iterative racing.

Param.	Description	Optimization range	Best
t	Tabu tenure	$\{0.5, 1, 1.5, 2, 2.5, 3\}p$	$0.5p$
I_{max}	Max. consecutive non-improving tabu iter.	$10^2 \times \{1, 5, 10, 25, 50\}$	500
A_{max}	Max. alternations between tabu searches	$\{1, 5, 10, 25, 50\}$	10
sp_{max}	Max. sp-escape iterations	$\{1, 5, 10, 25, 50\}$	25
α	Slack factor allowed to D when optimizing B	$[0.0, 0.2]$	0.0

and 32 GB of main memory running Ubuntu Linux 16.04. For each experiment, only one core was used.

The main parameters of our algorithms were calibrated with the irace package in GNU R [26], with a budget of 2000 runs and a time limit of 5 min per run. This time limit was chosen due to time availability reasons. To avoid overfitting, for the calibration we have generated an additional data set of 48 instances, with the same characteristics as DL. The parameter t was set relative to the number of districts p, as recommended by [6]. Table 1 shows the parameter ranges used for calibration, and the best values obtained by racing. Unexpectedly, the best value for α was found to be 0, which suggests that allowing the compactness to increase when optimizing balance is unhelpful.

In the following, we report experiments with instance set DL for several configurations. For each instance, we have executed 10 replications of the algorithm with 30 min of running time. For each instance size (n, p), we report averages over all replications of instances of that size. In each table, we report the average relative deviation of the diameters from the best known value "$D(\%)$", over replications that achieved feasibility (i.e., $B = 0$), and the empirical probability "p. f." of achieving feasibility on each multistart iteration, averaged over all replications.

4.3 Experiment 1: Constructive Algorithm

In Sect. 3.1 we argued that a construction strategy biased towards minimizing D is more effective, since optimizing balance during the local improvement phase is much easier than optimizing compactness. This experiment aims to support that argument by comparing our standard constructive approach (*greedy-D*) to two other strategies. The first one, *greedy-B*, greedily selects the best candidate c_i of each district with respect to balance B instead of diameter D. The second one, *BFS*, constructs solutions non-greedily by expanding free boundary-nodes in breadth-first order, each time assigning the expanded node to a neighboring district, with ties broken lexicographically. Both approaches generate initial seeds from the p-dispersion-based heuristic.

Table 2 shows the results of the three approaches. For each approach, columns B_C and $D_C(\%)$ report, for the iteration with smallest final diameter, the average imbalance and diameter (relative to the best known value) of the constructive part, whereas columns $D(\%)$ report the final average diameter (relative to the best known value) after local improvement.

Table 2. Comparison of constructive algorithms.

n	p	greedy-D				greedy-B				BFS			
		B_C	$D_C(\%)$	$D(\%)$	p. f.	B_C	$D_C(\%)$	$D(\%)$	p. f.	B_C	$D_C(\%)$	$D(\%)$	p. f.
1,000	5	0.19	48.97	**0.13**	0.99	0.22	51.51	0.52	1.00	3.09	64.07	0.18	1.00
1,000	10	0.97	65.08	**0.03**	0.89	0.86	105.63	2.45	0.93	6.71	122.12	0.10	0.92
1,000	16	2.67	104.91	**1.21**	0.43	1.90	134.61	6.46	0.58	12.13	160.74	1.96	0.56
2,500	12	1.27	63.33	**0.47**	0.99	1.05	99.32	11.12	0.99	8.39	118.86	1.28	1.00
2,500	25	3.14	102.58	**1.15**	0.77	3.42	137.02	16.21	0.83	17.48	218.61	2.62	0.86
2,500	40	7.97	115.68	**6.41**	0.21	6.17	187.62	21.06	0.30	28.61	246.94	9.07	0.32
5,000	25	2.25	88.98	**1.63**	0.97	2.46	136.12	28.40	0.98	15.73	164.05	3.33	0.99
5,000	50	8.17	112.37	**1.93**	0.58	7.62	236.50	34.81	0.67	32.48	266.82	5.89	0.74
5,000	80	16.46	95.33	**7.50**	0.05	12.90	240.87	43.48	0.09	56.92	316.57	12.53	0.12
10,000	50	6.89	106.42	**2.01**	0.92	8.77	146.55	48.14	0.94	29.83	249.46	7.40	0.94
10,000	100	17.73	109.97	**5.29**	0.34	14.81	264.57	51.79	0.46	60.90	307.61	11.64	0.49
10,000	160	34.86	120.40	**46.54**	0.01	28.28	220.67	93.79	0.01	107.97	301.49	69.26	0.02
Average		8.55	94.50	**6.19**	0.60	7.37	163.42	29.85	0.65	31.69	211.45	10.44	0.66

We can see that *greedy-D* leads to smallest diameters, on average, followed by *BFS* and *greedy-B*. As expected, *greedy-D* had the advantage in initial compactness and *greedy-B* in initial balance (though a small one), both yielding significantly better initial solutions than *BFS*. Yet, all three methods had a similar empirical probability (p. f.) of finding a feasible solution after local improvement, at each multistart iteration. This supports our claim that the difficulty to make near-balanced and highly imbalanced initial solutions feasible during local improvement is similar.

Although *BFS* has worse initial solutions, after local improvement they become competitive in both compactness and probability of finding a feasible solution. A possible explanation for this unexpected success is that, because instances in DL have very regular topologies (a Delaunay triangulation of uniform random points), a breadth-first strategy which, by design, typically yields compact districts in graph space will also yield somewhat compact districts in Euclidean space.

4.4   Experiment 2: Search Strategies

This experiment evaluates the effectiveness of the alternating tabu search strategy (A-TS), alternating local searches (A-LS) and a single tabu search with an objective function which is a weighted sum of balance and compactness (W-TS). For the weighted approach, we use the same objective function as [7]: $W(S) = 0.3\,B(S) + 0.7\,D(S)/d_{max}$, where $d_{max} = \max_{u,v \in V} d_{uv}$ is the maximum distance of two nodes. The weighted approach repeatedly performs a single tabu search in the neighborhood N_{bal} with the goal of minimizing $W(S)$, stopping after I_{max} consecutive non-improving iterations. As long as D does not change, we can use the candidate caching explained in Sect. 3.2 to avoid re-evaluating N_{bal} fully every time. For the local search approach, we stop as soon as no candidates improve the incumbent solution, but still allow a fixed A_{max} alternating searches.

Table 3. Comparison of search strategies.

n	p	A-TS			A-LS			W-TS		
		$D(\%)$	Iter.	p. f.	$D(\%)$	Iter.	p. f.	$D(\%)$	Iter.	p. f.
1,000	5	0.13	24,429	0.99	**0.08**	118,903	0.99	0.41	12,904	0.99
1,000	10	**0.03**	20,775	0.89	0.90	118,291	0.76	3.35	10,819	0.80
1,000	16	**1.21**	10,787	0.43	4.25	112,855	0.18	10.30	12,162	0.34
2,500	12	**0.47**	13,085	0.99	3.54	33,536	0.93	7.09	5,265	0.99
2,500	25	**1.15**	10,168	0.77	7.39	34,733	0.32	19.63	5,436	0.72
2,500	40	**6.41**	6,276	0.21	72.62	34,186	0.01	34.21	6,772	0.15
5,000	25	**1.63**	7,767	0.97	9.02	13,178	0.51	21.63	2,980	0.96
5,000	50	**1.93**	5,200	0.58	15.76	13,602	0.03	37.73	3,333	0.53
5,000	80	**7.50**	3,697	0.05	101.86	12,941	0.00	51.54	3,767	0.04
10,000	50	**2.01**	3,406	0.92	16.27	4,932	0.11	41.19	1,597	0.90
10,000	100	**5.29**	2,240	0.34	213.95	4,914	0.00	59.17	1,737	0.29
10,000	160	**46.54**	1,786	0.01	—	4,545	0.00	97.47	1,530	0.00
Average		**6.19**	9,135	0.60	40.51	42,218	0.32	31.98	5,692	0.56

Table 3 shows the results. Columns "Iter." show the average number of multistart iterations for each approach. All approaches use constructive heuristics *greedy-D*, and thus column "D" of A-TS is the same as for *greedy-D* in Table 2.

The experiments show that the alternating tabu search A-TS leads to the best results, with the smallest diameters, on average, and highest probability of finding a feasible solution. The alternating local search A-LS leads to a similar average diameter as the weighted tabu search W-TS, but is significantly worse than A-TS in both diameter and achieving feasibility, failing to solve the largest instance with $n = 10000$ and $p = 160$. This confirms that the tabu search is effective. Both W-TS and A-TS have a similar chance of finding a feasible solution, but A-TS finds significantly better diameters, on average. This indicates that, while a weighted approach does well at improving balance, an alternating strategy and techniques like the sp-escape play an important part in reducing the diameter. The number of iterations per second of A-LS is as expected the highest, since the search stops at the first local minimum, followed by A-TS and W-TS. The number of iterations of A-TS is higher for small p but decreases faster as p grows, while the number of iterations for A-LS and W-TS appears to be independent of p.

4.5   Experiment 3: Comparison with Existing Methods

This experiment compares our methods to the GRASP with path relinking (PR) proposed by [7] to solve the same problem. We refer the reader to their original paper for a more detailed description of their algorithms.

[7] have kindly made available to us all data sets and source code developed in their research. Because their implementation was done in MATLAB, we chose to reimplement their algorithms in C++ and compile them under the same environment as our own. In the reimplementation, we have used the "static" PR variant proposed by them, as it reportedly produces better results, on average. For a fair comparison, in the reimplementation we have used the optimization techniques we proposed in Sect. 3.4 to maintain diameters and connectivity dynamically. We have also adapted their PR algorithm, as we have found that it does not always maintain connectivity during the path relinking process: we simply stop the process once no more moves maintain connectivity. Further, we have found that ordering PR moves by expected increase to the objective function was more effective than the original approach, which orders moves lexicographically by node index. Both modifications have improved the overall performance. Finally, because the termination criterion of their algorithm is not a maximum time limit, but rather a fixed number of iterations, in our implementation we repeat their full algorithm while there is still time and, in the end, report the best intermediate solution.

Table 4 shows the results. Columns under "RME-16" refer to our implementation of [7]'s algorithm. As before, for each instance we report averages over 10 replications of 30 min of running time each. The first two rows show average results for data sets DS and DT, respectively, and the following for data set DS. We report the average number of multistart iterations of our method ("Iter."), and the number of times [7]'s method was repeated under the time limit ("Phases"), as well as the probability of the methods finding a feasible solution in a single iteration ("p. it.") or phase ("p. ph."). Note that these values are not directly comparable, since one refers to multistart iterations and the other to replications of the full algorithm, and the former are two orders of magnitude more than the latter in the same time.

We can see that our method yields significantly more compact solutions in every instance size of data set DL, with better overall results in data sets DS and DT as well. For DL, the algorithm of [7] usually had a high probability of finding

Table 4. Comparison to existing methods.

	n	p	Our method			RME-16		
			$D(\%)$	Iter.	p. it.	$D(\%)$	Phases	p. ph.
DS	500	10	**0.96**	23,112	0.29	7.46	375	0.98
DT	500	10	**0.03**	99,520	1.00	4.43	433	1.00
DL	1,000	5	**0.13**	24,429	0.99	1.08	96	1.00
	1,000	10	**0.03**	20,775	0.89	11.41	100	1.00
	1,000	16	**1.21**	10,787	0.43	22.07	87	1.00
	2,500	12	**0.47**	13,085	0.99	16.86	28	1.00
	2,500	25	**1.15**	10,168	0.77	34.00	23	1.00
	2,500	40	**6.41**	6,276	0.21	40.56	17	0.84
	5,000	25	**1.63**	7,767	0.97	38.91	7	1.00
	5,000	50	**1.93**	5,200	0.58	48.44	3	1.00
	5,000	80	**7.50**	3,697	0.05	109.27	2	0.14
	10,000	50	**2.01**	3,406	0.92	50.47	1	1.00
	10,000	100	**5.29**	2,240	0.34	78.57	1	0.50
	10,000	160	**46.54**	1,786	0.01	—	0	0.00
Average			**5.38**	16,589	0.60	35.66	84	0.82

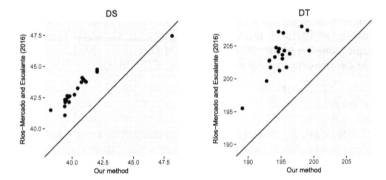

Fig. 3. Comparison of the final diameters achieved by our method and our implementation of [7]'s, for data set DS (left) and DT (right).

a feasible solution in a single phase, only failing to solve the largest instances with $n = 10000$ and $p = 160$. Our method had somewhat low probabilities of feasibility for DS as well as instances in DL with high p, but, given the sheer total number of iterations, it was always able to solve each instance at least once.

Figure 3 compares average final diameters obtained by both methods on each instance of data sets DS and DT. We can see that our method achieved lower values for all instances of data set DT, and for all but one instance (d500-03) of data set DS. In fact, this was the only case where their method consistently achieved better results in all replications. This could indicate that our method has a hard time exploring some parts of the search space, even in small instances.

5   Concluding Remarks

We have proposed a multistart heuristic that uses an alternating tabu search strategy to solve a commercial districting problem. Our method differs from previous approaches because it improves each optimization criterion separately through a customized tabu search, as opposed to using a composite fitness function that assigns weights to each objective. We have performed experiments on two data sets from the literature, and one proposed in this paper, with a wider range of instance sizes. The results confirmed the effectiveness of the different components of our method and showed that it has a high probability of finding feasible solutions of good quality. Our method is also competitive when compared to existing approaches in the literature, significantly improving state of the art upper bounds in almost all cases.

As a final note, we believe the proposed alternating search strategy can be applied without much change to other districting problems that model compactness as diameters, as [27], as well as related grouping problems such as the maximum dispersion problem [28], whose objective has a dual correspondence to the diameter.

Acknowledgments. This research was supported by the Brazilian funding agencies CNPq (grant 420348/2016-6), FAPEMIG (grant TEC-APQ-02694-16) and by Google Research Latin America (grant 25111). We would also like to thank to support of the Fundação de Desenvolvimento Científico e Cultural (FUNDECC/UFLA).

References

1. Ricca, F., Scozzari, A., Simeone, B.: Political districting: from classical models to recent approaches. Ann. Oper. Res. **204**(1), 271–299 (2013)
2. Ricca, F., Simeone, B.: Local search algorithms for political districting. Eur. J. Oper. Res. **189**(3), 1409–1426 (2008)
3. Bozkaya, B., Erkut, E., Haight, D., Laporte, G.: Designing new electoral districts for the city of Edmonton. Interfaces **41**(6), 534–547 (2011)
4. Bação, F., Lobo, V., Painho, M.: Applying genetic algorithms to zone design. Soft. Comput. **9**(5), 341–348 (2005)
5. Ríos-Mercado, R.Z., Fernández, E.: A reactive GRASP for a commercial territory design problem with multiple balancing requirements. Comput. Oper. Res. **36**(3), 755–776 (2009)
6. Lei, H., Laporte, G., Liu, Y., Zhang, T.: Dynamic design of sales territories. Comput. Oper. Res. **56**, 84–92 (2015)
7. Ríos-Mercado, R.Z., Escalante, H.J.: GRASP with path relinking for commercial districting. Exp. Syst. Appl. **44**, 102–113 (2016). (September 2015)
8. Camacho-Collados, M., Liberatore, F., Angulo, J.M.: A multi-criteria Police Districting Problem for the efficient and effective design of patrol sector. Eur. J. Oper. Res. **246**(2), 674–684 (2015)

9. Steiner, M.T.A., Datta, D., Steiner Neto, P.J., Scarpin, C.T., Rui Figueira, J.: Multi-objective optimization in partitioning the healthcare system of Parana State in Brazil. Omega **52**, 53–64 (2015)

10. Blais, M., Lapierre, S.D., Laporte, G.: Solving a home-care districting problem in an urban setting. J. Oper. Res. Soc. **54**(11), 1141–1147 (2003)

11. Gliesch, A., Ritt, M., Moreira, M.C.O.: A genetic algorithm for fair land allocation. In: Genetic and Evolutionary Computation Conference, pp. 793–800. ACM Press (2017)

12. Kalcsics, J.: Districting problems. In: Laporte, G., Nickel, S., da Gama, F.S. (eds.) Location Science, pp. 595–622. Springer, Cham (2015). https://doi.org/10.1007/978-3-319-13111-5_23

13. Salazar-Aguilar, M.A., Ríos-Mercado, R.Z., Cabrera-Ríos, M.: New models for commercial territory design. Netw. Spat. Econ. **11**(3), 487–507 (2011)

14. Salazar-Aguilar, M.A., Ríos-Mercado, R.Z., González-Velarde, J.L.: GRASP strategies for a bi-objective commercial territory design problem. J. Heuristics **19**(2), 179–200 (2013)

15. Feo, T.A., Resende, M.G.C.: A probabilistic heuristic for a computationally difficult set covering problem. Oper. Res. Lett. **8**(2), 67–71 (1989)

16. Butsch, A., Kalcsics, J., Laporte, G.: Districting for arc routing. INFORMS J. Comput. **26**(October), 809–824 (2014)

17. Glover, F.: Future paths for integer programming and links to artificial intelligence. Comput. Oper. Res. **13**, 533–549 (1986)

18. Erkut, E., Ülküsal, Y., Yeniçerioğlu, O.: A comparison of p-dispersion heuristics. Comput. Oper. Res. **21**(10), 1103–1113 (1994)

19. Tarjan, R.E.: A note on finding the bridges of a graph. Inf. Process. Lett. **2**(6), 160–161 (1974)

20. Łącki, J., Sankowski, P.: Optimal decremental connectivity in planar graphs. Theory Comput. Syst. **61**(4), 1037–1053 (2016)

21. King, D.M., Jacobson, S.H., Sewell, E.C., Cho, W.K.T.: Geo-graphs: an efficient model for enforcing contiguity and hole constraints in planar graph partitioning. Oper. Res. **60**(5), 1213–1228 (2012)

22. Shamos, M.I.: Computational Geometry. Ph.D. thesis (1978)

23. Har-Peled, S.: On the Expected Complexity of Random Convex Hulls, pp. 1–20, November 2011. http://arxiv.org/abs/1111.5340

24. Andrew, A.M.: Another efficient algorithm for convex hulls in two dimensions. Inf. Process. Lett. **9**(5), 216–219 (1979)

25. Overmars, M.H., van Leeuwen, J.: Maintenance of configurations in the plane. J. Comput. Syst. Sci. **23**(2), 166–204 (1981)

26. López-Ibáñez, M., Dubois-Lacoste, J., Pérez Cáceres, L., Birattari, M., Stützle, T.: The irace package: iterated racing for automatic algorithm configuration. Oper. Res. Perspect. **3**, 43–58 (2016)

27. Chou, C., Kimbrough, S.O., Sullivan-Fedock, J., Woodard, C.J., Murphy, F.H.: Using interactive evolutionary computation (IEC) with validated surrogate fitness functions for redistricting. In: Genetic and Evolutionary Computation Conference, pp. 1071–1078 (2012)

28. Fernández, E., Kalcsics, J., Nickel, S.: The maximum dispersion problem. Omega **41**(4), 721–730 (2013)

An Ant Colony Approach for the Winner Determination Problem

Abhishek Ray[1] and Mario Ventresca[2(✉)]

[1] Krannert School of Management, Purdue University,
West Lafayette, IN 47906, USA
ray52@purdue.edu
[2] School of Industrial Engineering, Purdue University,
West Lafayette, IN 47906, USA
mventresca@purdue.edu

Abstract. Combinatorial auctions are those where bidders can bid on bundles of items. These auctions can lead to more economically efficient allocations but determining the winners is an NP-complete problem. In this paper, we propose an ant colony technique for approximating solutions to hard instances of this problem. Hard instances are those that are unsolvable within reasonable time by CPLEX and have more than 1000 bids on 500 or more unique items. Such instances occur in real world applications such as 4th Party Logistics (4PL) auctions, online resource time sharing auctions and the sale of spectrum licenses by the Federal Communications Commission. We perform experiments on 10 such instances to show and compare the performance of the proposed approach to CPLEX (Branch-and-Bound), stochastic greedy search, random walk and a memetic algorithm. Results indicate that in a given runtime, CPLEX results lie within the third quartile of the values generated using our approach for 3 of 10 of the instances. In addition, CPLEX results are on average 0.24% worse than best values reported using our approach for 5 of 10 instances. Further, our approach performs statistically significantly better ($p < 0.01$) than other heuristics on all instances.

Keywords: Ant colony · Auctions · Winner determination

1 Introduction

A combinatorial auction is a type of smart market in which bidders can place bids on combinations of discrete items rather than on individual items. For example, in a radio spectrum auction for wireless communications, combinatorial auctions are becoming the mechanism of choice for efficient allocation of resources [1]. However, determining the winner in such auctions, so that the auctioneer's revenue is maximized is an NP-complete problem [1]. As the number of market participants and items increases, there is a need for computationally efficient

© Springer International Publishing AG, part of Springer Nature 2018
A. Liefooghe and M. López-Ibáñez (Eds.): EvoCOP 2018, LNCS 10782, pp. 174–188, 2018.
https://doi.org/10.1007/978-3-319-77449-7_12

methods for approximating the winners. Heuristics have been valuable for finding an approximate solution when classic methods fail or require excessive computation time. In this paper, we propose an ant colony-based metaheuristic for the winner determination problem (WDP) in combinatorial auctions (CA). We represent the WDP as a graphical formulation and search for optimal solutions. We conduct experiments on benchmark test instances to show the effectiveness of the metaheuristic compared to an exact approach (CPLEX) and other inexact approaches (memetic algorithm, stochastic greedy search).

1.1 Main Contributions

Our main contributions are as follows:

1. **Problem Formulation** - we devise a procedure to convert a set of bids and bid values into a Directed Acyclic Graphical (DAG) structure that preserves the feasibility of allocation. In contrast to past work [3], we formulate the bid allocation problem as a path-finding problem, where the longest path gives the maximum possible revenue for the auctioneer. This also has advantages in proving convergence results of the heuristic. Specifically, as outlined in [4], graph based ant colony formulations have the special property, under certain conditions, of converging close to the optimal with probability close to 1.
2. **Randomized Pheromone Updating** - we adapt the Max-Min (MMAS) approach as described in [5] for a better search of the solution landscape. In contrast to MMAS which considers either iteration best or global best ant, our approach randomizes pheromone updates between iteration best ant, global best ant and a mathematically derived fixed upper and lower bounds of pheromones. This improves the trade-off between 'explore' and 'exploit' when searching the landscape for the global optimal.
3. **Randomized Graph Pruning** - we devise a procedure that allows for fast convergence to approximate solutions. Specifically, by maintaining a list of the number of times an ant visits an edge we approximate how valuable the edge is to the solution. If it is not valuable, we discard the edge to reduce the search space.
4. **Solving Hard Instances of WDP** - past work using branch-and-bound approaches for solving WDP [6] used problem instances that consisted of 1000 bids from between 50 and 256 unique items. These instances were drawn from different prior distributions (e.g. uniform, decay) and were optimally solvable using CPLEX within 400 s, on average. Our approach is for those instances of WDP for which CPLEX fails to return a result within stipulated time. Specifically, CPLEX either fails to solve (running out of memory when solving MIPs is commonly reported issue with using branch and bound [28]) or takes unreasonably long to solve these instances. We use benchmark problem instances that were generated by [7]. These problem instances are realistic as they reflect bidder preference, pricing and bidding behavior more accurately than previous attempts such as [25]. These problem instances contain more than 1000 bids from between 500 and 1500 unique items. Our approach demonstrates the benefits of using metaheuristics on such instances.

The paper is organized as follows. Section 2 explains the winner determination problem and its formulation. Section 3 gives a brief overview of the past work in solving WDP. Subsequently, we explain our approach in Sects. 4, 5, 6. Finally, the experimental results are shown in Sect. 7.

2   Winner Determination Problem

The Winner Determination Problem (WDP) has the goal of finding winning bids that maximizes an auctioneers revenue under the constraint that each item can be allocated to at most one bidder. The mathematical formulation of WDP is equivalent to the weighted set packing problem [6]. Consider a set of items $S = \{1, 2, 3, 4..., k\}$ to be auctioned and a set of l bids, $B = \{b_1, b_2, ..., b_l\}$. A bid b_j is a tuple defined as (S_j, q_j) where $S_j \subseteq S$ is a subset of items and $q_j > 0$ is the maximum bid price associated with S_j. Further, consider a matrix $A_{k \times l}$ having k rows and l columns, such that $A_{ij} = 1$ if item i belongs to bid j. Finally, the decision variables are $x_j = 1$ if bid B_j is accepted as winning and $x_j = 0$ otherwise. The formulation for maximizing auctioneer's revenue is

$$\max_{x_j} \sum_{j=1}^{l} q_j x_j \text{ s.t. } \sum_{j=1}^{l} A_{ij} x_j \leq 1 \ \forall \ i \in \{1, 2, 3, 4...., k\} \tag{1}$$

Note that the constraints impose that each item can be allocated to at most one bidder. This formulation is provably NP-complete [9,15].

3   Literature Review

Approaches to solving WDP can be broadly classified into two categories - Exact Methods and Inexact Methods.

3.1   Exact Methods

Exact methods, given enough time, are guaranteed to find an optimal solution. The well-known exact algorithms for the WDP are based on Branch-and-Bound [8,9]. Other work has Branch-on-Items (BoI) [10], Branch-on-Bids (BoB) [11], and Combinatorial Auctions BoB (CABoB) [6]. These methods can find optimal allocation for instances containing hundreds of items. The CASS (Combinatorial Auction Structural Search) is a Branch-and-Bound algorithm for the WDP proposed by [12]. Further, [13] proposed CAMUS (Combinatorial Auctions Multi-Unit Search) which is a new version of the CASS for determining the optimal set of bids in general multi-unit combinatorial auctions. Improving on CASS, CAMUS introduces a novel branch-and-bound technique that makes use of several additional procedures such as - dynamic programming techniques to more efficiently handle multi-unit single-good bids.

There have been attempts at solving the WDP using other traditional approaches. For example, [14] used dynamic programming and [15] proposed another exact algorithm based on integer programming. [16] used constraint programming to solve the WDP when faced with bid withdrawal problem [27]. In general, exact algorithms perform well on test instances when the number of items is between 50 and 200. However, like most branch-and-bound techniques, solving certain instances can be time-consuming and the number of nodes in a branching tree can be computationally unmanageable.

3.2 Inexact Methods

Inexact methods, given enough time, may find optimal solutions, but they cannot be used to prove the optimality of any solution they find. In general, inexact methods are based on heuristics or metaheuristics and they are helpful for finding good solutions for very large instances of WDP (e.g., instances with bids between 1000 to 1500 bids). These instances occur regularly in industry in shipping route and resource scheduling auctions [7]. Generally, such auctions have between 500 to 1000 unique items that make exact approaches slow and at times, inefficient.

The current well-known inexact algorithms for the WDP are: Hybrid Simulated Annealing SAGII [17], Casanova [18], stochastic local search [19] and memetic algorithms [20]. Overall, the memetic algorithm approach has been shown to outperform SAGII, Casanova and stochastic local search for finding solutions within comparable CPU time for a specific set of hard instances containing 1000–1500 bids [20]. The memetic algorithm approach to solving WDP combines diversification and intensification of search in a way that leads to better results. However, even with this approach, stochastic local search is used to better explore the search space, which indicates that there is still a need to develop better heuristics for WDP. Our approach aims to fill this gap in literature.

4   Proposed Approach

4.1   Preprocessing Phase

Our formulation consists of two steps. The first step preprocesses the data of bundles and associated bids into a 2-D matrix with unique bundles and its corresponding maximum bid amount. For each bundle (defined as a subset of items up for auction), its corresponding maximum bid amount is determined and an $m \times n$ matrix is constructed, where m is the number of unique bundles and n is cardinality of the largest bundle. The reason for this preprocessing step follows from the problem definition for winner determination. Specifically, if several bids have been submitted on the same combination of items, for winner determination using the bid with the highest price maximizes auctioneer's revenue [6]. Other bids can be discarded as irrelevant since it can never be beneficial for the auctioneer to accept one of these inferior bids. Once the $m \times n$ matrix B is initialized, a directed graph $G(V, E)$ is constructed using the fact that any possible solution $S_{opt} \subseteq S$ for WDP should consist of bids such that if $S_i, S_j \in S_{opt}$, then it must be that $S_i \cap S_j = \emptyset$. Specifically, this matrix is used to construct a DAG as per the following procedure:

1. Let $G(V, E)$ have vertices as the unique bundles.
2. A directed edge exists between any two vertices iff the subsets are disjoint in nature. Specifically,

$$E = \{(S_i, S_j) \mid S_i \cap S_j = \emptyset \; \forall \; i, j \in V\}. \tag{2}$$

3. Add source vertex s and target vertex t to the vertex set (i.e. $V = V \cup \{s, t\}$). These vertices enforce a starting point and an ending point for a path traversed by each ant.
4. Create a directed edge from s to every vertex in $V \setminus \{t\}$. Similarly, create an edge from every vertex in $V \setminus \{s\}$ to t. Any edge from a vertex $V \setminus \{s, t\}$ to t is an outgoing edge from that vertex. No edge exists between s and t.
5. Vertices $V \setminus \{s, t\}$ are connected using two edges with different weights. The weight of the edge directed from v_i to v_j is the maximum bid value associated with bundle v_j and vice versa for edge directed from v_j to v_i. For edges from any vertex $V \setminus \{t\}$ to t the edge weight is a constant equal to zero.

$$w(e_{ij}) = q_j \; \forall \; e_{ij} \in E \text{ where } q_j = 0 \text{ if } j = \{t\}. \tag{3}$$

Example: Consider three unique bundles ($\{A\}, \{B\}, \{A, B\}$) and the associated maximum bid amounts for these bundles to be 3,4,9, respectively. Evidently, bundle $\{A, B\}$ is superior to other bundles due to associated bid 9. First, denote the vertices as $V = (s, \{A\}, \{B\}, \{A, B\}, t)$. Then construct the edges following procedure outlined above, as depicted in the Fig. 1 below. An example path on the graph is $s \to A \to B \to t$. The auctioneer gets a revenue of 7.

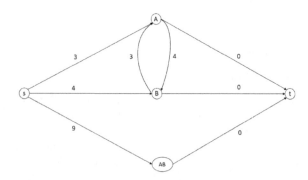

Fig. 1. Graph $G(V, E)$ from bundles as vertices and max bid amount as edges.

Once bundles and bids are translated into a graph with a definite start and end node, the ant system is executed. In addition, since edges are directed and acyclic, for any path constructed on the graph the ants can visit each node at most once. Algorithm 1 pseudocode of how the ants traverse the graph.

Algorithm 1. Ant System for WDP

Procedure: GBAS for WDP
1 **Input** - DAG $G(V, E)$
2 **Initialization** - No. of ants (N), Pheromone values τ_0, maximum
   iterations MaxIter, Solution set constructed by ant $S_{a_i} \forall i = 1, 2, 3..N$,
   Best Solution constructed S_a^{Best}
3 **Output** - S_a^{Best} - Best solution constructed by Ants
4 **while** $k \leq MaxIter$ **do**
5   Position ants $a_i \forall i = 1, 2, 3...N$ at vertex $v_i = s$
6   Initialize Path $= \emptyset$
7   **for** *Each ant* $a_i \in N$ *until vertex* t *is reached* **do**
8   Choose vertex v_j from v_i using $p_{ij} = \frac{\tau_{ij}^\alpha [w_{ij}]^\beta}{\sum_{j \in N(i)} \tau_{ij}^\alpha [w_{ij}]^\beta}$.
9   Path$=$ Path $\cup \, v_j$
10         $S_{a_i} := S_{a_i} + w(e_{ij})$
11         $v_i = v_j$
12     $S_a^{Best} = \max \left(S_a^{Best}, S_{a_i} \right)$
13   Pheromone evaporation on $G(V, E)$: $\tau'_{ij} = (1 - \rho)\tau_{ij}$
14   Pheromone update (Min-Max) using: $\tau_{ij}^{new} = \tau'_{ij} + \delta(S_a^{Best})$
15 **Return** S_a^{Best}
16 **End** Procedure

4.2 Theoretical Convergence to Optimal for WDP

Graph based ant system formulations can be theoretically shown to converge
on an optimal path of the graph [4], under certain assumptions. Proof of such
convergence has been shown for TSP formulations [23]. In this section we
prove the convergence to optimal for the WDP formulation. First, the following
conditions and notations are assumed for the convergence to be demonstrated.

1. Let $\alpha = 1$, where α is the weight given to pheromone in any general ACO
   implementation. By assuming $\alpha = 1$, magnifying the impact of increase or
   decrease in pheromone levels is relaxed. This implies that the local decision
   probability reduces to

$$p_{ij} = \frac{\tau_{ij}[w_{ij}]^\beta}{\sum_{j \in N(i)} \tau_{ij}[w_{ij}]^\beta}. \tag{4}$$

   This preserves pheromone values on the choice of each node by an ant and
   simplifies the analysis (similar to earlier proofs of convergence in [4]). This
   equation refers to vertex choice part of Algorithm 1.
2. Let the set of paths constructed on the graph during iteration t be denoted
   as $M(t)$. The iteration best path is denoted by $m_{Best}(t) \in M(t)$. The global
   optimal path is denoted by M_{opt}. An optimal path would start at s, end at t
   and contains each node of $V = \{v_1, v_2, v_3, ..., v_m\}$ at most once.

3. Every edge on the optimal path has a weight that is non-negative and equal to the bid value of node v_j. That is, $w_{ij} \geq 0 \; \forall$ edges $i, j \in M_{opt}$.
4. Solutions constructed by ants that are at least as good as the previous best solution get rewarded with increased pheromone.

We now outline the terminology and assumptions to be used in the convergence proof. Let L denote the set of edges on the optimal path M_{opt} and let $|L|$ be the number of edges in the path M_{opt}. Let there be N ants $(a_1, a_2, a_3 ..., a_N)$ traversing the graph in each iteration $t = \{1, 2, 3 ...\}$. An ant a_k at node i, will choose a node j depending on the probability $p_{ij}(t)$:

$$p_{ij}(t) = \frac{\tau_{ij}(t)[w_{ij}]^\beta}{\sum_{j \in N(i)} \tau_{ij}(t)[w_{ij}]^\beta} \tag{5}$$

We assume two conditions for the weights and pheromones on edges for simplicity of exposition. First, let the weights w_{ij} on the edges of the graph be normalized such that $w_{min} = \min\left(\{w_{ij}\}^\beta\right) \geq 0$ and $w_{max} = \max\left(\{w_{ij}\}^\beta\right) \leq 1$. Second, let pheromones τ_{ij} be normalized such that $\sum_{i,j \in m(t)} \tau_{ij}(t) [w_{ij}(t)]^\beta \leq 1$.

Denote E_t^k as the event that some ant $a_k \; \forall \; k = \{1, 2, 3 ... N\}$ traverses the optimal path m^* and $H_t = \overline{E_t^1} \wedge \overline{E_t^2} \wedge \overline{E_t^3} \wedge \overline{E_t^4} \wedge .... \wedge \overline{E_t^N}$ as the event that no ant traverses the optimal path at iteration t.

Proposition 1. *The probability $P(\overline{H_t}) = 1 - P(H_t)$ that at least one ant traverses the optimal path during iteration t is $P(\overline{H_t}) \geq 1 - (1 - \eta^L \hat{\rho}^{t-1})^N$, where $\hat{\rho} = (1 - \rho)^L$ and $\eta = w_{min}\tau_{i,j}(1)$.*

Proof. This is a proof by induction. The basic idea is to show that as the number of iterations $t \to \infty$, the probability that the ant system converges onto M_{opt} converges to 1.

Consider the pheromone update equations as described at the end of ant traversal loop in Algorithm 1. Restating here for clarity we have, $\tau_{ij}^{\text{new}}(t + 1) = \tau'_{ij}(t) + \delta(m_{\text{Best}}(t))$ which can be written as $\tau_{ij}(t+1) = (1-\rho)\tau_{ij}(t) + \delta(m_{\text{Best}}(t))$. Since $\delta(m_{\text{Best}}(t)) > 0$, we can directly conclude that $\tau_{ij}(t + 1) \geq (1 - \rho)\tau_{ij}(t)$. Since this holds true for all $t = \{1, 2, 3, ...\}$ we have that

$$\tau_{ij}(t) \geq (1 - \rho)^{t-1}\tau_{ij}(1). \tag{6}$$

The assumption of normalization of edge weights and $\sum_{i,j \in m(t)} \tau_{ij}(t) [w_{ij}(t)]^\beta \leq 1$, implies that for $p_{ij}(t)$ (from Eq. 5) at iteration t,

$$p_{ij}(t) \geq \tau_{ij}(t)[w_{ij}]^\beta \tag{7}$$

Now consider an optimal path of length $|L|$, where $L = \{s, v_r, v_{r+1}, ..., v_{r+L-2}, t\}$. Suppose an ant travels this optimal path in iteration t. Then by definition of event E_t^k we have,

$$P(E_t^k) = \prod_{(i,j)\in L} p_{i,j}(t) = p_{s,v_r}.p_{v_r,v_{r+1}}\cdots p_{v_{r+L-2},t} \tag{8}$$

$$\geq \prod_{(i,j)\in L} \tau_{ij}(t)[w_{ij}]^\beta \quad \text{(From Eq. 7)} \tag{9}$$

$$\geq w_{min}^L \prod_{(i,j)\in L} \tau_{ij}(t) \quad \text{(By normalization of edge weights)} \tag{10}$$

$$\geq w_{min}^L \prod_{(i,j)\in L} (1-\rho)^{t-1}\tau_{ij}(1) \quad \text{(From Eq. 6)} \tag{11}$$

$$\geq \left((1-\rho)^{(t-1)}\right)^L (w_{min}\tau_{ij}(1))^L \tag{12}$$

Let $\hat{\rho} = (1-\rho)^L$ and $\eta = w_{min}\tau_{ij}(1)$ then,

$$P(E_t^k) \geq (\hat{\rho})^{t-1}\eta^L. \tag{13}$$

This lower bound of $P(E_t^k)$ is derived for one ant. Since each ant independently traverses the graph,

$$P(\overline{E_t^1} \wedge \overline{E_t^2} \wedge \overline{E_t^3} \wedge \dots \wedge \overline{E_t^N}) = P(\overline{E_t^1})P(\overline{E_t^2})P(\overline{E_t^3})\dots P(\overline{E_t^N}). \tag{14}$$

Noting that $P(H_t) = P(\overline{E_t^1} \wedge \overline{E_t^2} \wedge \overline{E_t^3} \wedge \dots \wedge \overline{E_t^N})$,

$$P(H_t) \leq (1 - (\hat{\rho})^{t-1}\eta^L)^N \tag{15}$$

Therefore, $P(\overline{H_t}) = 1 - P(H_t) \geq 1 - (1 - (\hat{\rho})^{t-1}\eta^L)^N$ $\qquad\square$

Corollary 1. *When* $\alpha = 2$, $P(\overline{H_t}) = 1 - P(H_t) \geq 1 - (1 - (\hat{\rho})^{t-1}\eta^L)^N$ *where* $\hat{\rho} = (1-\rho)^{2L}$ *and* $\eta = w_{min}(\tau_{ij}(1))^2$.

Proof. From normalization of edge weights assumption, Eq. 7 still holds as $p_{ij}(t) \geq (\tau_{ij}(t))^2[w_{ij}]^\beta$. Since $\tau_{ij}(t) > 0$, Eq. 6 gives $\tau_{ij}(t) \geq (1-\rho)^{t-1}\tau_{ij}(1) \Rightarrow (\tau_{ij}(t))^2 \geq (1-\rho)^{2(t-1)}(\tau_{ij}(1))^2$. Substituting in expression for $P(E_t^k)$, the final probability expression simplifies to $P(\overline{H_t}) = 1 - P(H_t) \geq 1 - (1 - (\hat{\rho})^{t-1}\eta^L)^N$ where $\hat{\rho} = (1-\rho)^{2L}$ and $\eta = w_{min}(\tau_{ij}(1))^2$ $\qquad\square$

5 Randomized Pheromone Updating

We utilize a mechanism of pheromone update that probabilistically applies in three ways (1) either the global best ant solution (2) or the local best ant solution or (3) a mathematically derived fixed value of max and min pheromone. This strategy extends the Min-Max algorithm [24]. In MMAS, the strategy is to use only the global best or the (local) iteration best ant for pheromone update. However, in the proposed approach pheromone update is randomized between iteration best ant, global best ant and a mathematically derived fixed upper and lower bounds of pheromones. Using this approach, the likelihood that an ant gets stuck at a local optima is reduced [18]. Deciding the minimum and maximum level of pheromones is accomplished as outlined in Sect. 5.1.

5.1   Min-Max Pheromone Level

Pheromone updates are governed by the following rules:

$$\tau_{ij}(t) = (1 - \rho)\tau_{ij}(t), \ \forall \ \tau_{ij} \in m(t) \ \text{(Pheromone evaporation)} \tag{16}$$

$$\tau_{ij}(t+1) = (1 - \rho)\tau_{ij}(t) + \delta(m_{\text{Best}}(t)) \ \text{(Pheromone Deposit)} \tag{17}$$

Proposition 2. *For any* τ_{ij}, $\lim\limits_{t \to \infty} \tau_{ij}(t) \leq \tau_{max} = \frac{\delta(M_{opt})}{\rho}$ *where* M_{opt} *is the global optimal path that ants converge on.*

Proof. This is a proof by induction. The basic idea is that as $t \to \infty$, if the evaporation factor $0 < 1 - \rho < 1$ then the pheromone values asymptotically converge to a constant value τ_{max}.

At any iteration t, the pheromone deposit on any edge is

$$\tau_{ij}(t+1) = \tau_{ij}(t) + \delta(m_{\text{Best}}(t)) \tag{18}$$

Consider $t = 1$. The equations for pheromone update is,

$$\tau_{ij}(1) = (1 - \rho)\tau_{ij}(1) \tag{19}$$

$$\tau_{ij}(2) = (1 - \rho)\tau_{ij}(1) + \delta(m_{\text{Best}}(1)) \tag{20}$$

Where $\delta(m_{\text{Best}}(1))$ is pheromone update due to the optimal solution constructed during the first iteration by the best ant. Similarly, for the second iteration, we can generate similar update equations and simplify to:

$$\tau_{ij}(3) = (1 - \rho)\tau_{ij}(2) + \delta(m_{\text{Best}}(2)) \tag{21}$$

$$\Rightarrow (1 - \rho)\left((1 - \rho)\tau_{ij}(1) + \delta(m_{\text{Best}}(1))\right) + \delta(m_{\text{Best}}(2)) \tag{22}$$

$$\Rightarrow (1 - \rho)^2 \tau_{ij}(1) + (1 - \rho)\delta(m_{\text{Best}}(1)) + \delta(m_{\text{Best}}(2)) \tag{23}$$

Extending to t time periods, we can write the general expression as,

$$\tau_{ij}(t) = (1-\rho)^{t-1}\tau_{ij}(1) + (\delta(m_{\text{Best}}(t-1)) + (1 - \rho)\delta(m_{\text{Best}}(t-2))$$
$$+ (1 - \rho)^2 \delta(m_{\text{Best}}(t-3)) + ... + (1 - \rho)^{t-2}\delta(m_{\text{Best}}(1))) \tag{24}$$

In order to find the upper limit, the RHS needs to be maximal. We know that in this solution construction process, ants construct progressively higher valued solutions and hence, $\delta(m_{\text{Best}}(t)) \geq \delta(m_{\text{Best}}(t-1)) \geq \delta(m_{\text{Best}}(t-2)) \geq ... \geq \delta(m_{\text{Best}}(1))$, without loss of generality. Therefore, to determine τ_{max}, assuming $\delta(m_{\text{Best}}(t)) = \delta(M_{opt}) \ \forall \ t = 1, 2, 3...$ and simplifying we have,

$$\tau_{ij}(t) = (1 - \rho)^{t-1}\tau_{ij}(1) + \delta(m^*)\left(1 + (1 - \rho) + (1 - \rho)^2 + ... + (1 - \rho)^{t-2}\right) \tag{25}$$

Taking the limits as $t \to \infty$ and noting that $0 < (1 - \rho) < 1$;

$$\lim_{t \to \infty} \tau_{ij}(t) = \frac{\delta(M_{opt})}{1 - (1 - \rho)} = \frac{\delta(M_{opt})}{\rho} \tag{26}$$

which is the upper limit of pheromone levels.   □

τ_{max} restricts the value of pheromones to prevent convergence to a local optimal. As explained in [24], in general it has been empirically observed that when searching the space for good solutions there is a reasonable chance to find even better solutions in the area near good solutions. Hence, τ_{min} should not be too low and should vary with the quality of solutions found. Hence we assume that

$$\tau_{min} = \frac{\tau_{max}}{|L_t|} \tag{27}$$

where $|L_t|$ is the number of edges on the best path constructed by ants till iteration t. As explained in [24], the choice of τ_{min} is dependent on the designer and can be changed to vary search quality.

6   Randomized Graph Pruning

Randomized Graph Pruning (RGP) is employed to facilitate fast convergence of the ant system for those problem instances that are very large and have cardinality above 1000. The main benefit from RGP is efficient and fast approximations of solutions for those problem instances where exact approaches (such as Branch-and-bound as in CPLEX solvers) fails to provide quality approximations within a given timeframe. The pseudocode is shown in Algorithm 2.

Algorithm 2. Randomized Graph Pruning

Procedure: RGP for Graph Based Ant System

1 **Input** - $G(V, E)$, Counter matrix C of dimension $(|V| - 2) \times (|V| - 2)$,
   Iteration instances $t_c \subset t$ at which pruning is done
2 **Output** - Set of edges pruned E_c
3 **for** $k \in t_c$ **do**
4 ⎢   Initialize best path till iteration $k : L_k$
5 ⎢   Remove edges connected to best path edges from pruning
   ⎢   consideration $C(|L_k|, v_j) = 0 \; \forall \; v_j \subseteq V$, $C(v_i, |L_k|) = 0 \; \forall \; v_i \subseteq V$
6 ⎢   Construct edge set $F = \left\{ (v_i, v_j) : C(v_i, v_j) = \left[f \left(\frac{(t_c) \times (t_c + 1)}{2} \right) \right] \right\}$
7 ⎢ $E = E \setminus E_c$ where $E_c \subseteq F$ is randomly selected and $|E_c| = n$
8 **End** Procedure

In addition to the basic pheromone matrix, RGP uses a count matrix C of size $(|V| - 2) \times (|V| - 2)$. This matrix maintains count of the number of times ants traverse an edge and is initialized as

$$C_{ij} = \begin{cases} 1 \; \forall \; (v_i, v_j) \in G(V, E) \\ 0 \quad \text{otherwise} \end{cases} \tag{28}$$

The default value of each edge on the graph can be initialized to any non-negative number to keep count. For the current formulation, the default value is taken as 1. The basic idea behind prune count c_{ij} is as follows:

1. Each time an ant traverses an edge e_{ij}, $C_{ij} = C_{ij} + 1$.
2. Since ants choose edges probabilistically, the edges on best path in each iteration have a count c_{ij}^{max} proportional to the number of ants and number of iterations t. Conversely, edges not on best path during each iteration have an edge count $1 < c_{ij} < c_{ij}^{max}$.
3. Since edges connected to nodes on best path could be chosen in future iterations, these are not considered for pruning. $C(|L_k|, v_j) = 0 \ \forall \ v_j \subseteq V$, $C(v_i, |L_k|) = 0 \ \forall \ v_i \subseteq V$ refers to this.
4. For all other edges, those with count $1 < c_{ij} < c_{ij}^{max}$ are considered for pruning. This is for two reasons. First, $c_{ij} = 1$ would correspond to those edges with relatively small amount of pheromone that are unlikely to be chosen for future iterations and hence pruning them would not affect search. Second, those edges with values numerically close to c_{ij}^{max} should not be pruned since these have been traversed by ants in previous iterations and are possible solution candidates for future iterations.
5. The measure for prune count is $c_{ij} = \left\lceil f\left(\frac{(t_c) \times (t_c+1)}{2}\right)\right\rceil$ where t_c are iteration instances during which pruning is done (e.g., for $t = 1000$ iterations, pruning is done at $t_c = \{200, 400, 600, 800\}$ instance). Iteration instances are empirically determined. These are instances during iteration where the ants appear to be converging to a local optimal. The pruning changes the graph and hence, prevents convergence.
6. Function for approximating prune count is any slowly increasing concave function (e.g., logarithmic function). This is to ensure that as number of iterations increase, those edges with $c_{ij} \leq c_{ij}^{max}$ are not pruned.

7   Experimental Results

The experiments were run on an Intel i5-Core 3.5 GHz computer, with 16 GB RAM. The source code for preprocessing was written in R and for the ant colony system was written in MATLAB. For the ant system both RGP and randomized pheromone update were implemented. The ant system parameters α, β were empirically determined to be $\alpha = 2, \beta = 1.5$. Number of iterations for ant system in each trial run is 1500. The starting pheromone value on the graph is $\tau_0 = 1$ and pheromone evaporation factor is $\rho = 0.05$. Experiments were conducted on 10 out of the 20 problem instances used in [19] that were unsolvable within 601 s by CPLEX. This is the time taken by the ant system to construct a solution. These instances were initially developed in [7] and have been since used in [17,20,26]. Comparison is also performed with stochastic greedy search, random walk and memetic algorithm results from literature [19]. For stochastic greedy search and random walk, t-tests are performed to test whether there exists difference not due to randomness. For memetic algorithm, our results are compared with best reported results in [20]. The number of trials is 30 with an ant population size of 400 for each trial and problem instance. For 400 ants, average time taken to solve is 601 s and so we terminate CPLEX output at time 601 s as well. A box-plot of the results is shown to give an overall picture of the comparison with CPLEX and other heuristics. For the box-plots note that X-axis is test instances, Y-axis is solution value.

Table 1. Experimental trial result comparison table

Instance	CPLEX estimate (601 s)	Random walk best value	Stochastic greedy search best value	Memetic algorithm best value	ACO Median (400 ants)	ACO Max (400 ants)	% Difference
in101	67101.34	59183.26	61665.21	67101.93	**67809.65**	72724.62	8.38 %
in102	72518.56	54231.65	61796.54	67797.61	*69544.69*	71336.17	−1.63 %
in103	70263.26	55532.21	63287.12	66350.99	*67195.26*	69023.43	−1.10 %
in104	70951.21	53260.27	61135.32	64618.41	*70750.49*	72709.65	2.48 %
in105	71852.45	53930.91	62782.28	66376.83	*68063.42*	70774.45	−1.50 %
in106	66621.32	54418.94	61097.32	65481.64	*65359.59*	66604.63	−0.02 %
in107	69182.33	51999.89	61186.49	66245.70	*66418.32*	69398.19	0.31 %
in108	75147.12	58960.96	68008.55	74588.51	72393.98	74588.51	−0.74 %
in109	66439.67	50400.76	58287.17	62492.66	*64277.28*	68652.18	3.33 %
in110	65735.34	55823.04	59770.63	65171.19	*65460.89*	68295.29	4.72 %

Table 1 lists the results from our results as compared to CPLEX and other heuristics. The last column lists the percentage difference between ACO Max and CPLEX values. Statistically significant results ($p < 0.01$) (compared to CPLEX best value or memetic algorithm best values) is highlighted. If median is statistically significantly better than CPLEX then the number is bold, if better than memetic algorithm result then the number is italicized and if better than both then both bold and italicized. Median of results statistically significantly exceed CPLEX results within stipulated CPU run time of 601 s, for instance

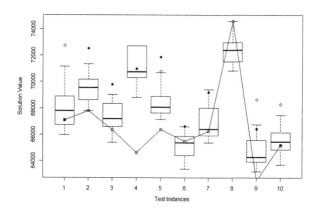

Fig. 2. Comparison with CPLEX and memetic algorithm results. Circled dots are CPLEX values. Square boxes connected through lines maximum values from memetic algorithm.

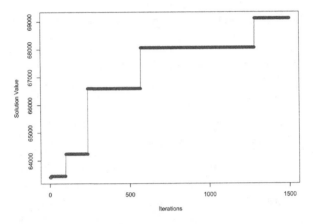

Fig. 3. Typical convergence pattern for 400 ant simulation run for 1500 iterations and randomized graph pruning happening at $t_c = \{200, 450, 700, 950, 1350\}$.

in101. CPLEX result lies within third quartile of our results for instances in101, in104, in110. Further, CPLEX result is on average 3.28% is not as good as the maximum value of our results for instances in101, in104, in107, in109, in110. The time taken by memetic algorithm to give its best value is 172 s for a starting population size of 300. Our approach takes 601 s for a starting ant population size of 400. For an increase of 429 s in time taken for generating solutions, our approach produces results that exceed the best memetic algorithm values by, 6% on average. For example, for instance intance in104 and in109, best results from our approach are 13% and 10% higher than best memetic algorithm results. For all test instances except instance 8, best memetic algorithm results lie within third quartile of results from our approach (Fig. 2). In addition, for all instances but 8, best memetic algorithm results are statistically significantly lesser than median of our results. The maximum value from our approach is on average 6% higher than best memetic algorithm result for all instances. Results from our approach are statistically significantly higher than results from stochastic greedy search and random walk trial results.

Figure 3 shows the typical convergence pattern for our approach. The stepwise increase in objective value coincides with randomized graph pruning happening at iteration instances $t_c = \{200, 450, 700, 950, 1350\}$. These instances are empirically determined to be the instances at which pruning should be executed.

8   Conclusion and Future Research

We use ant colony optimization to solve hard instances of the winner determination problem. Randomized pheromone updating and randomized graph pruning were implemented to increase the speed of search. We test our approach on hard instances of the problem and compare the results with CPLEX, memetic algorithm, stochastic greedy search and random walk. Results indicate that in CPLEX results lie within third quartile of our results for 3 of 10 instances. Further, for 5 of 10 instances producing, CPLEX is on average lesser by 3.28%. In comparison to the memetic algorithm approach, our approach gives median values that are statistically significantly higher than memetic algorithm best values. Further, the maximum values from our approach is on average 6% higher than best memetic algorithm results for all instances. Although results are encouraging, the aim is to further extend the approach to other difficult instances. We also plan to test our heuristic with the other metaheuristics as outlined in [19] and refine the heuristic to better scale for instances having more than 1500.

References

1. Bichler, M., Gupta, A., Ketter, W.: Research commentary - designing smart markets. Inf. Syst. Res. **21**(4), 688–699 (2010)
2. Dorigo, M., Süttzle, T.: The ant colony optimization metaheuristic: algorithms, applications, and advances. In: Glover, F., Kochenberger, G.A. (eds.) Handbook of Metaheuristics. International Series in Operations Research & Management Science, vol. 57, pp. 251–286. Springer, Boston (2003). https://doi.org/10.1007/0-306-48056-5_9
3. Gan, R., Guo, Q., Chang, H., Yi, Y.: Ant colony optimization for winner determination in combinatorial auctions. In: Third International Conference on Natural Computation, ICNC 2007, vol. 4, pp. 441–445. IEEE, August 2007
4. Gutjahr, W.J.: A graph-based ant system and its convergence. Future Gener. Comput. Syst. **16**(8), 873–888 (2000)
5. Stutzle, T., Hoos, H.: MAX-MIN ant system and local search for the traveling salesman problem. In: IEEE International Conference on Evolutionary Computation, pp. 309–314. IEEE, April 1997
6. Sandholm, T., Suri, S., Gilpin, A., Levine, D.: CABOB: a fast optimal algorithm for winner determination in combinatorial auctions. Manag. Sci. **51**(3), 374–390 (2005)
7. Lau, H.C., Goh, Y.G.: An intelligent brokering system to support multi-agent Web-based 4/sup th/-party logistics. In: Proceedings of 14th IEEE International Conference on Tools with Artificial Intelligence, ICTAI 2002, pp. 154–161. IEEE (2002)
8. Sandholm, T.: Algorithm for optimal winner determination in combinatorial auctions. Artif. Intell. **135**(1–2), 1–54 (2002)
9. Sandholm, T., Suri, S., Gilpin, A., Levine, D.: Winner determination in combinatorial auction generalizations. In: Proceedings of the First International Joint Conference on Autonomous Agents and Multiagent Systems: Part 1, pp. 69–76. ACM, July 2002
10. Sandholm, T., Suri, S.: Improved algorithms for optimal winner determination in combinatorial auctions and generalizations. In: AAAI/IAAI, pp. 90–97, July 2000
11. Sandholm, T., Suri, S.: BOB: improved winner determination in combinatorial auctions and generalizations. Artif. Intell. **145**(1–2), 33–58 (2003)
12. Fujishima, Y., Leyton-Brown, K., Shoham, Y.: Taming the computational complexity of combinatorial auctions: optimal and approximate approaches. In: IJCAI, vol. 99, pp. 548–553, July 1999

13. Leyton-Brown, K., Shoham, Y., Tennenholtz, M.: An algorithm for multi-unit combinatorial auctions. In: AAAI/IAAI pp. 56–61, July 2000
14. Rothkopf, M.H., Peke, A., Harstad, R.M.: Computationally manageable combinational auctions. Manag. Sci. **44**(8), 1131–1147 (1998)
15. Andersson, A., Tenhunen, M., Ygge, F.: Integer programming for combinatorial auction winner determination. In: Proceedings of Fourth International Conference on MultiAgent Systems, pp. 39–46. IEEE (2000)
16. Holland, A., O'Sullivan, B.: Robust solutions for combinatorial auctions. In: Proceedings of the 6th ACM Conference on Electronic Commerce, pp. 183–192. ACM, June 2005
17. Guo, Y., Lim, A., Rodrigues, B., Zhu, Y.: Heuristics for a bidding problem. Comput. Oper. Res. **33**(8), 2179–2188 (2006)
18. Hoos, H.H., Boutilier, C.: Solving combinatorial auctions using stochastic local search. In: AAAI/IAAI, pp. 22–29, July 2000
19. Boughaci, D., Benhamou, B., Drias, H.: Local search methods for the optimal winner determination problem in combinatorial auctions. J. Math. Model. Algorithms **9**(2), 165–180 (2010)
20. Boughaci, D., Benhamou, B., Drias, H.: A memetic algorithm for the optimal winner determination problem. Soft Comput.-Fusion Found. Methodol. Appl. **13**(8), 905–917 (2009)
21. Dorigo, M., Maniezzo, V., Colorni, A., Maniezzo, V.: Positive feedback as a search strategy (1991)
22. Dorigo, M., Di Caro, G.: Ant colony optimization: a new meta-heuristic. In: Proceedings of the 1999 Congress on Evolutionary Computation, CEC 1999, vol. 2, pp. 1470–1477. IEEE (1999)
23. Gutjahr, W.J.: A generalized convergence result for the graph-based ant system metaheuristic. Probab. Eng. Inf. Sci. **17**(4), 545–569 (2003)
24. Sttzle, T., Hoos, H.H.: MAX-MIN ant system. Future Gener. Comput. Syst. **16**(8), 889–914 (2000)
25. Leyton-Brown, K., Pearson, M., Shoham, Y.: Towards a universal test suite for combinatorial auction algorithms. In: Proceedings of the 2nd ACM Conference on Electronic Commerce, pp. 66–76. ACM, October 2000
26. Alidaee, B., Kochenberger, G., Lewis, K., Lewis, M., Wang, H.: A new approach for modeling and solving set packing problems. Eur. J. Oper. Res. **186**(2), 504–512 (2008)
27. Porter, D.P.: The effect of bid withdrawal in a multi-object auction. Rev. Econ. Des. **4**(1), 73–97 (1999)
28. Wilson, D.G., Rudin, B.D.: Introduction to the IBM optimization subroutine library. IBM Syst. J. **31**(1), 4–10 (1992)

Erratum to: On the Fractal Nature
of Local Optima Networks

Sarah L. Thomson, Sébastien Verel, Gabriela Ochoa,
Nadarajen Veerapen, and Paul McMenemy

Erratum to:
Chapter "On the Fractal Nature of Local Optima Networks"
in: A. Liefooghe and M. López-Ibáñez (Eds.):
Evolutionary Computation in Combinatorial Optimization,
LNCS 10782, https://doi.org/10.1007/978-3-319-77449-7_2

In the original version of the paper, Figure 4 had been re-generated incorrectly during a camera-ready modification and reflected a different dataset. Therefore, some of the correlations, density plots, and scatter-plots displayed in Figure 4 did not reflect the results discussed in the paper. The revised paper contains the corrected figure showing the results as they should be.

The updated online version of this chapter can be found at
https://doi.org/10.1007/978-3-319-77449-7_2

Author Index

Printed in the United States
By Bookmasters